T0134538

# Springer Theses

Recognizing Outstanding Ph.D. Research

## Aims and Scope

The series "Springer Theses" brings together a selection of the very best Ph.D. theses from around the world and across the physical sciences. Nominated and endorsed by two recognized specialists, each published volume has been selected for its scientific excellence and the high impact of its contents for the pertinent field of research. For greater accessibility to non-specialists, the published versions include an extended introduction, as well as a foreword by the student's supervisor explaining the special relevance of the work for the field. As a whole, the series will provide a valuable resource both for newcomers to the research fields described, and for other scientists seeking detailed background information on special questions. Finally, it provides an accredited documentation of the valuable contributions made by today's younger generation of scientists.

## Theses are accepted into the series by invited nomination only and must fulfill all of the following criteria

- They must be written in good English.
- The topic should fall within the confines of Chemistry, Physics, Earth Sciences, Engineering and related interdisciplinary fields such as Materials, Nanoscience, Chemical Engineering, Complex Systems and Biophysics.
- The work reported in the thesis must represent a significant scientific advance.
- If the thesis includes previously published material, permission to reproduce this must be gained from the respective copyright holder.
- They must have been examined and passed during the 12 months prior to nomination.
- Each thesis should include a foreword by the supervisor outlining the significance of its content.
- The theses should have a clearly defined structure including an introduction accessible to scientists not expert in that particular field.

More information about this series at http://www.springer.com/series/8790

Felix Winterstein

# Separation Logic for High-level Synthesis

Doctoral Thesis accepted by
Imperial College London, UK

*Author*
Dr. Felix Winterstein
Department of Electrical
and Electronic Engineering
Imperial College London
London
UK

*Supervisor*
Prof. George A. Constantinides
Imperial College London
London
UK

ISSN 2190-5053 ISSN 2190-5061 (electronic)
Springer Theses
ISBN 978-3-319-85094-8 ISBN 978-3-319-53222-6 (eBook)
DOI 10.1007/978-3-319-53222-6


Softcover reprint of the hardcover 1st edition 2017


Printed on acid-free paper

This Springer imprint is published by Springer Nature
The registered company is Springer International Publishing AG
The registered company address is: Gewerbestrasse 11, 6330 Cham, Switzerland

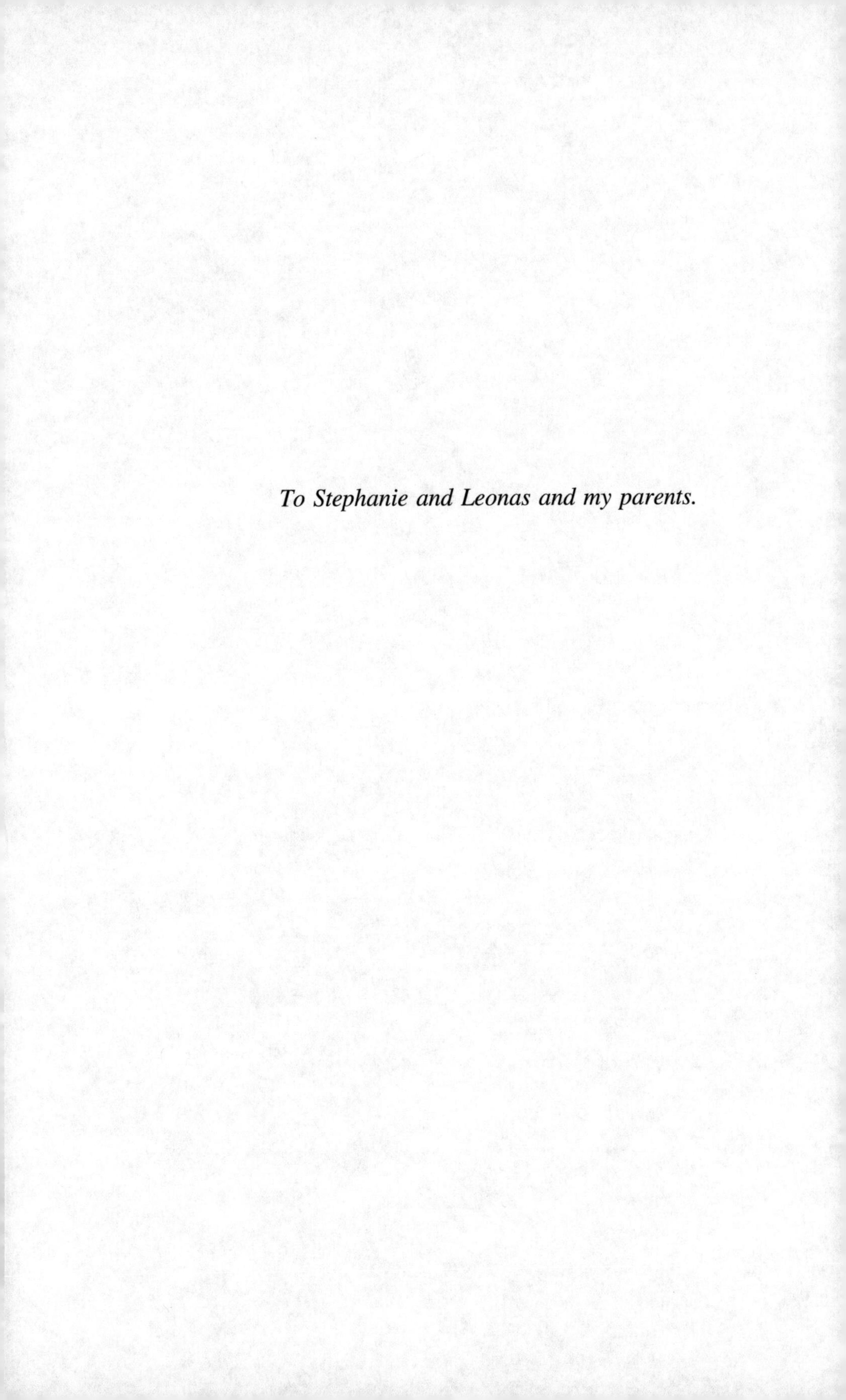

*To Stephanie and Leonas and my parents.*

# Supervisor's Foreword

It gives me great pleasure to introduce this book, based on the work of my former Ph.D. student Felix Winterstein. I have worked in the field of High-Level Synthesis (HLS) for more than 15 years, and it has never been as exciting as it is today. Industry is beginning to adopt HLS as the design flow of choice. As a result, research that once ended up purely as academic papers now finds itself pored over by a variety of industry vendors, all looking to improve their HLS offering. At the same time, the adoption of FPGAs in the datacentre is driving the demand for—and excitement about—FPGA HLS. This book contains the core ideas for a radical extension of the reach of HLS to encompass efficient hardware realisations of a wide variety of algorithms previously out of reach to existing design tools. As with all the best academic research, there is a significant departure from the industrial status quo here, yet Dr. Winterstein has integrated his approach with existing design tools so that the promise of the approach is visible—and accessible—to all. In these pages, you will find a skilful and intelligent blend of mathematical logic, algorithms, digital design and compilers. I hope you enjoy reading this book as much as I enjoyed supervising the work.

November 2016

Prof. George A. Constantinides
Imperial College London
London
UK

# Abstract

High-level synthesis (HLS) promises a significant shortening of the digital hardware design cycle by raising the abstraction level of the design entry to high-level languages such as C/C++. However, applications using dynamic, pointer-based data structures remain difficult to implement well, yet such constructs are widely used in software. Automated optimisations that leverage the memory bandwidth of dedicated hardware implementations by distributing the application data over separate on-chip memories and parallelise the implementation are often ineffective in the presence of dynamic data structures, due to the lack of an automated analysis that disambiguates pointer-based memory accesses. This thesis takes a step towards closing this gap. We explore recent advances in *separation logic*, a rigorous mathematical framework that enables formal reasoning about the memory access of heap-manipulating programs. We develop a static analysis that automatically splits heap-allocated data structures into provably disjoint regions. Our algorithm focuses on dynamic data structures accessed in loops and is accompanied by automated source-to-source transformations which enable loop parallelisation and physical memory partitioning by off-the-shelf HLS tools.

We then extend the scope of our technique to pointer-based memory-intensive implementations that require access to an off-chip memory. The extended HLS design aid generates parallel on-chip multi-cache architectures. It uses the disjointness property of memory accesses to support non-overlapping memory regions by private caches. It also identifies regions that are shared after parallelisation and that are supported by parallel caches with a coherency mechanism and synchronisation, resulting in automatically specialised memory systems. We show up to 15× acceleration from heap partitioning, parallelisation and the insertion of the custom cache system in demonstrably practical applications.

# Abstract

# Acknowledgements

I would like to thank my Ph.D. advisor Prof. George Constantinides for his encouragement, for his excellent support and for sparking my interest in separation logic. George has given me the freedom to pursue my own research questions and helped guide my thoughts with sage advice, sharp insight and inspiring ideas. It is largely due to the research environment he has created that this work resulted in a thesis of which I am proud.

I thank my colleagues at Imperial, especially Shane Fleming, Andrea Suardi, John Wickerson, Samuel Bayliss, Michalis Vavouras, Grigorios Mingas, Nadesh Ramanathan, David Boland and Ivan Beretta who have made the Circuits and Systems Group an inspiring and vibrant work environment. Special thanks go to Shane and his wife Iza who have become close friends and regularly hosted me during the many short visits to London.

I also thank the European Space Agency for funding my research and my former colleagues from ESA's Ground Station Systems Division for being such pleasant company. In particular, I would like to thank Gunther Sessler for being an excellent mentor and a friend. In addition to him, it was with the help of Marco Lanucara and Piermario Besso that my application for ESA's contribution to my Ph.D. funding was successful.

Outside of Imperial and ESA, I would like to thank Hsin-Jung Yang from MIT and Elliott Fleming from Intel for a very productive collaboration during the last two years.

I am grateful for the advice and support given by my parents Sabine and Bernhard Winterstein and my brother Florian.

Finally, I thank my wonderful partner Stephanie for her love, patience and support.

You made this thesis possible.

# Acknowledgements

# Contents

# List of Figures

# List of Tables

# Chapter 1
# Introduction

With the increasing demand for performance and efficiency of computing devices, *custom computing* is a growing area in digital computation today, which represents a class of processing devices that are dedicated to an application or a range of similar applications. Custom computing devices can achieve higher energy or power efficiency and performance with respect to general-purpose microprocessors, which can execute any task on the same underlying hardware [1–9]. Efficiency and performance are gained by avoiding unnecessary circuitry for a specific computing task, and the design of custom data paths and memory systems. The trade-off between flexibility and performance/efficiency varies across different classes of specialised computing machines: Digital signal processors and application-specific instruction set processors (ASIPs) are software-programmable and provide extended hardware support for domain-specific features. On the other hand, digital application-specific integrated circuits (ASICs) are fully customised processors that implement computation based on a digital circuit which is usually dedicated to a single application; once produced, the functionality of an ASIC is hard-wired and cannot be changed. Field-programmable gate arrays (FPGAs) have a particular role in the flexibility/performance trade-off in that they combine programmability with an efficient dedicated circuit implementation for a particular application. An FPGA consists of configurable logic cells and interconnects and typically can be reprogrammed to implement different computing tasks post-fabrication.

The traditional design entries of ASIC and FPGA implementations are largely similar in the first phases of the design flow. A hardware model is written in a hardware description language (HDL) such as VHDL [10] or Verilog [11] at the level of abstraction referred to as register transfer level (RTL). The specification at RTL allows the user to have full control over the low-level details of the data path and memory system implementations on the chip and to navigate the implementation through a large design space. However, producing a manual RTL specification requires significant design and verification effort, including several iterations of design optimisation and verification phases. The development times for complex ASIC implementations may amount to several years until tape-out, while RTL design and verification dominate the overall development cycle. The design cycle for FPGA implementations is



F. Winterstein, *Separation Logic for High-level Synthesis*, Springer Thesis,
DOI 10.1007/978-3-319-53222-6_1

typically shorter, but the design effort at RTL is similar. Long implementation cycles are a hindrance for an adoption of FPGAs as efficient yet flexible processing devices: reprogrammability encourages their use in a similar way as microprocessors are used in that the same hardware can execute different 'programs'. However, prohibitively long development times compared to software implementations fundamentally limit this versatility. Furthermore, the RTL design entry inevitably requires familiarity with the low-level details of digital hardware design. The conceptual difference between the application development for FPGAs and for instruction set architectures hinders the wide adoption of FPGA technology by software developers and application engineers without experience in circuit design [12].

The low productivity of application development at RTL has encouraged the electronic design automation (EDA) community to raise the abstraction level of application descriptions from RTL to high-level languages such as C/C++. High-level synthesis (HLS) tools take these descriptions as input and automatically generate RTL specifications which can be synthesised and mapped into hardware by standard back-end RTL tool flows. High-level design entry can significantly shorten the development cycle when compared with RTL-based specification. Remarkable effort in academia and industry has led to various HLS tools targeting ASIC and FPGA technology. With C/C++ being one of the most prevalent programming languages used to date and with large bases of legacy codes written in it, RTL compilation from C/C++ and derivatives thereof has a long-standing tradition in industrial [13–22] and academic [23–27] development. The admissible source code entry to these tools is restricted to synthesisable subsets of the C language.

HLS has experienced an increased interest in the last decade, which we believe is due to two main reasons. Firstly, state-of-the-art tools have increased performance compared to previous generations of tools developed in the mid 1990s [28]. The performance of an HLS tool can be measured in the quality of results (QoR) of the resulting RTL description in terms of execution time and hardware resource utilisation. Recent evaluations [29–31] show that state-of-the-art tools, such as Xilinx Vivado HLS [20], can achieve a QoR comparable to hand-written HDL code. Secondly, technology scaling has brought the number of transistors on a chip to point where the RTL design effort required to make efficient use this resource is becoming an increasingly severe limitation [12]. On the other hand, the abundance of hardware resources makes the trade-off between the QoR of hand-written HDL and generated HLS designs and design times appealing to more and more users, a fact that is especially true for FPGA implementations whose end-to-end development time is usually significantly shorter than that of ASIC designs [12, 32].

Despite the encouraging QoR results of FPGA-targeted HLS evaluations for particular benchmarks [29–33], there are types of programs that either cannot be synthesised at all, or result in a poor QoR. Applications using dynamic, pointer-based data structures and dynamic memory allocation are examples of such programs. The objective of this thesis is to extend the scope of current HLS to such *pointer-based* programs. Our work is motivated by the fact that pointer-based memory references and dynamic memory allocation are well established and widely used features of high-level languages such as C++. However, their analysis and automated program optimisations resulting from it are beyond the scope of the overwhelming majority of HLS techniques to date. Although dynamic memory allocation, an unsupported feature in common HLS flows, can be made synthesisable with manual source code modifications, pointer-based programs often do not result in efficient hardware implementations. As we shall see in Chap. 2, the HLS implementation of such a program can be degraded by a factor larger than 26× in terms of execution time compared to a hand-crafted RTL design if the source code is not further optimised prior to HLS. The reward for extensive manual code optimisations is shown to be an 8× improvement of the execution time.

We identify two aspects that are crucial for improving the QoR. The first is the extraction of parallelism from a pointer-based application while preserving the program semantics, which is usually based on a dependence analysis. Secondly, computational parallelism requires that the memory system is not a sequential bottleneck to performance. We aim to make efficient use of the customisable memory architecture in FPGAs, which is a key feature distinguishing FPGAs from microprocessors. Instead of a monolithic memory space, the application data can be distributed over many small blocks of on-chip memory leading to a high aggregate memory bandwidth. Consequently, multiple computational units can be fed in parallel which results in a very efficient parallelisation if expensive dynamic interconnects between any memory and any worker in a parallel computational unit can be reduced to single peer-to-peer connections, i.e. the parallelism is communication-free. The C model, however, assumes the presence of a *heap*, a large monolithic memory space in which a program allocates and frees up portions at run time. The difficulty of parallelisation and memory partitioning lies in the disambiguation of memory references: regardless of scope, every two heap-directed pointers potentially alias, i.e. reference the same memory cell, which leads to dependencies between expressions that are syntactically unrelated. The difficulty of analysing these programs is exacerbated by linked data structures which contain pointers in their link fields.

Expanding on the encouraging results in Chap. 2, the scope of this thesis is to automate source code transformations that enable parallelisation and memory partitioning in HLS flows. We present a static program analysis which breaks the monolithic heap memory into several disjoint portions, which we refer to as *heaplets* in this thesis, and rules out dependencies between code fragments that a standard HLS tool must assume potentially exist. The dependence/disjointness analysis enables automated source-to-source transformations for parallelisation and data distribution which can be exploited by a back-end HLS tool. Our departure point from previous work is the use of recent advances in *separation logic* [34], a mathematical framework that allows a rigorous formal description of the program state and reasoning about the resources accessed by a program. Separation logic extends the classical propositional logic by an operator that explicitly expresses the separation of resources, i.e. the non-aliasing property of two pointers. This paves the way for an automated program analysis and can straightforwardly handle dynamic memory allocation in disjoint heaplets. Separation logic has predominantly been leveraged in modern software verification tools. To the best of our knowledge, its application in the context of automated code optimisations for HLS remains largely unexplored. Experiments in Chap. 4, comparing the automatically parallelised to the direct HLS implementations, show an average reduction of execution time by a factor of 2.4× across several benchmarks.

Besides the on-chip memory partitioning and parallelisation, our source-to-source transformations ensure the synthesisability of heap-manipulating programs including dynamic memory allocation by standard HLS tools. The implementations in Chap. 4 are constructed under the assumption that the application data fits in the physical on-chip memory. However, the chances of exhausting the memory resources in an FPGA application with a large memory footprint are high since the maximum capacity of on-chip memory in state-of-the-art FPGAs is only in the order of tens of megabytes. We remove the limitation of being restricted to on-chip memory implementations in Chap. 5 by embedding HLS kernels in a framework that provides access to an external memory hierarchy consisting of board-level dynamic random access memory (DRAM) and host machine-level main memory. Accessing external memory, however, can substantially slow down the FPGA accelerator due to memory bandwidth limitations and, in the worst case, the contention on the external memory bus eliminates the gain of parallelisation. The starting point for our work in Chap. 5 is the insertion of on-chip caches to buffer frequently reused data and to reduce the number of expensive accesses to the external memory.

Our main contribution in Chap. 5 is the application of an extended version of the baseline analysis in Chap. 4 to the automatic generation of an application-specific on-chip multi-cache architecture. Firstly, we extend the analysis such that, at compile time, it provides precise information about which regions in heap memory will be shared after the implementation has been parallelised. This extends its scope to programs whose memory access pattern does not allow a partitioning into fully independent computational units and therefore broadens the applicability of our technique. Secondly, we use the disjointness/sharing information to instantiate an application-specific, hybrid multi-cache system that contains *private* caches for heap regions known to be private for a computational unit and caches with an additional (and inherently more expensive) *coherence mechanism* and synchronisation service for shared heap regions. In the remainder of this thesis, we distinguish between these two modes by referring to *private* and *coherent* caches, while the latter case corresponds to inter-cache coherency. We also extend the multi-cache construction with a technique for custom sizing so as to maximise the aggregate hit rate in private caches under a memory resource constraint. We demonstrate a speed-up of up to 15.2× after parallelisation and generation of a multi-cache architecture compared to the unparallelised application and uncached access to the off-chip memory. Furthermore, the hybrid system outperforms a default all-coherent version by 69.3% on average in terms of the area-time product across our benchmarks.

This thesis moves us towards the goal of supporting full-featured C/C++ code in future HLS flows by providing a framework that enables efficient FPGA acceleration of irregular computation over pointer-based data structures. In Sect. 4.3, we propose an approach to integrate this framework into future HLS tools. The overall vision is that 'standard' software codes, including those from legacy code bases which have not been developed with HLS in mind, can be equally seamlessly mapped to FPGA accelerators while leaving the platform-specific optimisations to the compiler. This further raises the level of abstraction in digital hardware design and may lead to a wider adoption of FPGA technology in an extended scope of applications.

## 1.1 Research Contributions

This thesis makes the following main contributions:

- A separation logic-based parallelisation algorithm for pointer-based programs which access dynamic data structures. Our static program analysis handles straight-line code as well as arbitrary `while`-loops and determines whether communication-free parallelism can be exposed in the loop execution with respect to the accessed dynamic data structures. Starting from the C memory model of a global monolithic heap memory, it determines how to partition the heap and dynamic data structures into disjoint partitions that can be implemented in separate on-chip memory blocks.
- The implementation of an automated source-to-source transformation infrastructure: The source translator ensures synthesisability of code containing unsupported constructs related to dynamic memory allocation. In a second pass, the disjointness information provided by our analysis is used to split the synthesised heap memory into separate blocks and to split a loop into multiple loops so as to obtain a semantically equivalent parallel implementation. The property of communication-free parallelism ensures that each functional unit only requires access to its own private memory block.
- In addition to the identification of disjoint heap regions, we extend the baseline heap analysis by an identification of heaplets that would be shared by the parallel loop kernels after parallelisation. Our analysis inserts additional synchronisation primitives for program fragments that access shared resources. Even if coherency is ensured, updates to the shared resource may happen in a different order after parallelisation compared to the sequential program. We present a *commutativity analysis* for the shared heap update in order to prove that the parallelisation is semantics-preserving.
- The extended framework targets FPGA accelerators with access to an off-chip memory. The disjointness and sharing information provided by our analyses are used to break the heap (residing in off-chip memory by default) into heaplets, to generate an application-specific parallel multi-scratchpad architecture containing on-chip caches and (if needed) coherency mechanisms: we synthesise parallel private scratchpads for disjoint heap regions and (inherently more expensive) coherent parallel scratchpads for shared regions.
- We further extend this framework by automated size scaling of private on-chip caches that uses spare on-chip memory resources. We generate individual sizing information for the multi-cache system and find the best size distribution for a user-provided memory access pattern of a particular application.

## 1.2 Thesis Outline

Before discussing the background and related work on program analyses, parallelisation and memory system optimisations in an HLS context in Chap. 3, this thesis begins with the presentation of a case study in the next chapter. The case study compares RTL and HLS implementations of two alternative algorithms for the same compute-intensive machine learning application (clustering) with significantly different computational properties: a data-flow centric implementation and a recursive tree traversal implementation that incorporates data-dependent control flow and makes use of pointer-linked data structures and dynamic memory allocation. The reason for this order of Chaps. 2 and 3 is two-fold: (1) It introduces the type of problems this work addresses and provides a motivating example for mapping an efficient pointer-based algorithm to an FPGA rather than its pointer-less brute-force counterpart. (2) It shows the capabilities and limitations of an exemplary state-of-the-art C-to-FPGA tool when synthesising pointer-based programs and proposes a set of manual source code alterations that result in a significantly more efficient HLS design.

Chapter 3, after the discussion of related work, introduces separation logic, the theoretical framework that provides the foundation of our program analyses in Chaps. 4 and 5. The analysis in Chap. 4 automates an important part of the code transformations of Chap. 2 that enables memory partitioning and parallelisation. Chapter 5 extends the scope of this work to the construction of multi-cache systems and shared memory accesses. Chapter 6 concludes this thesis and summarises the key ideas and concepts developed in this work. It also outlines directions of future research that build on the research contributions made in this thesis.

## 1.3 Statement of Originality

This thesis is my own work and all related work is appropriately referenced. The original contributions made in this thesis have been published in the following peer-reviewed conference papers and journal articles:

1. F. Winterstein, S. Bayliss and G.A. Constantinides, "Separation Logic for High-Level Synthesis," *ACM Transactions on Reconfigurable Technology and Systems (TRETS)*, vol. 9, no. 2, pp. 10:1–10:23, Dec. 2015 [35].
2. F. Winterstein, K. Fleming, H.-J. Yang, J. Wickerson, G. Constantinides, "Custom-Sized Caches in Application-Specific Memory Hierarchies," *Proceedings of the IEEE International Conference on Field-Programmable Technology (ICFPT)*, pp. 144–151, 2015 [36].
3. F. Winterstein, K. Fleming, H.-J. Yang, S. Bayliss, G. Constantinides, "MATCHUP: Memory Abstractions for Heap Manipulating Programs," *Proceedings of the ACM/SIGDA International Symposium on Field-Programmable Gate Arrays (FPGA)*, pp. 136–145, 2015 [37].

4. F. Winterstein, S. Bayliss, G. Constantinides: "Separation Logic-Assisted Code Transformations for Efficient High-Level Synthesis," *Proceedings of the IEEE International Symposium on Field-Programmable Custom Computing Machines (FCCM)*, pp. 1–8, 2014 (best paper nominee) [38].
5. F. Winterstein, S. Bayliss, G. Constantinides: "High-Level Synthesis of Dynamic Data Structures: A Case Study Using Vivado HLS," *Proceedings of the IEEE International Conference on Field-Programmable Technology (ICFPT)*, pp. 362–365, 2013 [31].
6. F. Winterstein, S. Bayliss, G. Constantinides: "FPGA-based K-means Clustering Using Tree-Based Data Structures," *Proceedings of the International Conference on Field Programmable Logic and Applications (FPL)*, pp. 1–6, 2013 [39].

The C-based HLS and RTL source code developed for the case study in Chap. 2 were made publicly available in an open source repository[1] [40].

Our work on cache architecture specialisation uses the open-source LEAP (Latency-insensitive Environment for Application Programming) framework [41] to embed the C/C++-based HLS kernels in an environment that constructs on-chip caches and an interface to external DRAM and host system main memory. LEAP is developed jointly at the Massachusetts Institute of Technology (MIT, Computer Science and Artificial Intelligence Laboratory) and the Intel Software and Services Group. The work in Chap. 5 and the corresponding publications [35, 37] were done in collaboration with the LEAP developers Kermin Elliott Fleming from Intel and Hsin-Jung Yang from MIT. Their main contribution was support for integrating our HLS kernels in the LEAP environment. Furthermore, following discussions about automatic cache scaling (also presented in Chap. 5), they implemented a new cache micro-architecture in LEAP that uses buffered banks of on-chip memory to support higher clock rates in large caches, an implementation that is used by our technique. In turn, our HLS benchmarks have been used to support the cache architecture design space explorations, which has led to my co-authorship in the following joint publications:

1. H.-J. Yang, K. Fleming, M. Adler, F. Winterstein, J. Emer, "LMC: Automatic Resource-Aware Program-Optimized Memory Partitioning," *Proceedings of the ACM/SIGDA International Symposium on Field-Programmable Gate Arrays (FPGA)*, pp. 128–137, 2016 [42].
2. H.-J. Yang, K. Fleming, M. Adler, F. Winterstein, J. Emer, "Scavenger: Automating the Construction of Application-Optimized Memory Hierarchies," *Proceedings of the IEEE International Conference on Field Programmable Logic and Applications (FPL)*, pp. 1–8, 2015 [43].

The collaboration with Intel/MIT also resulted in a tutorial session jointly held at the International Conference on Field Programmable Logic and Applications (FPL) in 2015 [44]. Some of the HLS, RTL and Bluespec System Verilog source code

[1] https://github.com/FelixWinterstein/Vivado-KMeans.

developed within the scope of Chap. 5 was also made publicly available in an open source repository[2] [45].

Finally, the RTL and HLS implementations developed in the scope of Chap. 2 have been included in other research projects (a case study for dynamic load balancing on FPGAs, fault mitigation in an FPGA-based space processor, and a hardware compiler for higher order functional programs). My contribution to these projects resulted in a co-authorship of the following publications:

1. N. Ramanathan, J. Wickerson, F. Winterstein, G.A. Constantinides, "A Case for Work-stealing on FPGAs with OpenCL Atomics," *Proceedings of the ACM/SIGDA International Symposium on Field-Programmable Gate Arrays (FPGA)*, pp. 48–53, 2016 [46].
2. S.T. Fleming, D.B. Thomas, F. Winterstein, *FPGAs and Parallel Architectures for Aerospace Applications: Soft Errors and Fault-Tolerant Design*. Springer International Publishing, 2016, ch. "A Power-Aware Adaptive FDIR Framework Using Heterogeneous System-on-Chip Modules", pp. 75–90 [47].
3. E.A. Pelaez, S. Bayliss, A. Smith, F. Winterstein, D.R. Ghica, D. Thomas, G.A. Constantinides: "Compiling Higher Order Functional Programs to Composable Digital Hardware," *Proceedings of the IEEE International Symposium on Field-Programmable Custom Computing Machines (FCCM)*, pp. 234–234, 2014 [48].

## References


1. J. Fowers, G. Brown, P. Cooke, G. Stitt, A performance and energy comparison of fpgas, gpus, and multicores for sliding-window applications," in *Proceedings of the ACM/SIGDA International Symposium on Field Programmable Gate Arrays (FPGA)* (2012), pp. 47–56
2. B. Cope, P.Y.K. Cheung, W. Luk, L. Howes, Performance comparison of graphics processors to reconfigurable logic: a case study. IEEE Trans. Comput. **59**(4), 433–448 (2010)
3. A. Putnam, A. Caulfield, E. Chung, D. Chiou, K. Constantinides, J. Demme, H. Esmaeilzadeh, J. Fowers, G. Gopal, J. Gray, M. Haselman, S. Hauck, S. Heil, A. Hormati, J.-Y. Kim, S. Lanka, J. Larus, E. Peterson, S. Pope, A. Smith, J. Thong, P. Xiao, D. Burger, A reconfigurable fabric for accelerating large-scale datacenter services, in *Proceedings of the ACM/IEEE International Symposium on Computer Architecture (ISCA)* (2014), pp. 13–24
4. Z.K. Baker, M.B. Gokhale, J.L. Tripp, Matched Filter Computation on FPGA, Cell and GPU, in *Proceedings of the IEEE International Symposium on Field-Programmable Custom Computing Machines (FCCM)* (2007), pp. 207–218
5. H. Riebler, T. Kenter, C. Plessl, C. Sorge, Reconstructing AES key schedules from decayed memory with FPGAs, in *Proceedings of the IEEE International Symposium on Field-Programmable Custom Computing Machines (FCCM)* (2014), pp. 222–229
6. J. Chase, B. Nelson, J. Bodily, Z. Wei, D.J. Lee, Real-time optical flow calculations on FPGA and GPU architectures: a comparison study, in *Proceedings of the IEEE International Symposium on Field-Programmable Custom Computing Machines (FCCM)* (2008), pp. 173–182
7. S. Asano, T. Maruyama, Y. Yamaguchi, Performance comparison of FPGA, GPU and CPU in image processing, in *Proceedings of the International Conference on Field Programmable Logic and Applications (FPL)* (2009), pp. 126–131


[2]https://github.com/FelixWinterstein/LEAP-HLS.

8. H. Park, S. Vijayvargiya, A. DeHon, Energy minimization in the time-space continuum, in *Proceedings of the International Conference on Field Programmable Technology (ICFPT)* (2015), pp. 64–71
9. J. Qiu, J. Wang, S. Yao, K. Guo, B. Li, E. Zhou, J. Yu, T. Tang, N. Xu, S. Song, Y. Wang, H. Yang, Going deeper with embedded FPGA platform for convolutional neural network, in *Proceedings of the 2016 ACM/SIGDA International Symposium on Field-Programmable Gate Arrays* (2016), pp. 26–35
10. IEEE Standard VHDL Language Reference Manual, *IEEE Std 1076-2008 (Revision of IEEE Std 1076-2002)* (2009), pp. 1–626
11. IEEE Standard for Verilog Hardware Description Language, *IEEE Std 1364-2005 (Revision of IEEE Std 1364-2001)* (2006), pp. 1–560
12. D. Bacon, R. Rabbah, S. Shukla, FPGA programming for the masses. Queue **11**(2), 40:40–40:52 (2013)
13. Y Explorations Excite, Accessed 01 Mar 2016. http://www.yxi.com/products.php
14. Calypto Catapult Synthesis, Accessed 19 Dec 2015. http://calypto.com/en/products/catapult/overview/
15. Cadence C-to-Silicon Compiler, Accessed 23 Dec 2015. http://www.cadence.com/products/sd/silicon_compiler/
16. Cadence Cynthesizer Solution, Accessed 23 Dec 2015. http://www.cadence.com/products/sd/cynthesizer/
17. Cadence Stratus High-Level Synthesis, Accessed 22 Dec 2015. http://www.cadence.com/products/sd/stratus/
18. Impulse CoDeveloper, Accessed 25 Nov 2015. http://www.impulseaccelerated.com/products.htm
19. Synopsys Synphony C Compiler, Accessed 26 Nov 2015. https://www.synopsys.com/Tools/Implementation/RTLSynthesis/Pages/SynphonyC-Compiler.aspx
20. Xilinx Vivado HLS, Accessed 12 May 2015. http://www.xilinx.com/products/design-tools/vivado/integration/esl-design.html
21. Altera SDK for OpenCL, Accessed 13 Jan 2016. https://www.altera.com/products/design-software/embedded-software-developers/opencl/overview.html
22. Xilinx SDAccel Development Environment for OpenCL, Accessed 13 Jan 2016. http://www.xilinx.com/products/design-tools/software-zone/sdaccel.html
23. High-Level Synthesis with LegUp, Accessed 20 Oct 2015. http://legup.eecg.utoronto.ca/
24. C. Pilato, F. Ferrandi, Bambu: a modular framework for the high level synthesis of memory-intensive applications, in *Proceedings International Conference on Field Programmable Logic and Applications (FPL)* (2013), pp. 1–4
25. ROCCC 2.0|Jacquard Computing, Accessed 12 May 2015. http://www.jacquardcomputing.com/roccc/
26. R. Nane, V. M. Sima, B. Olivier, R. Meeuws, Y. Yankova, K. Bertels, DWARV 2.0: A CoSy-based C-to-VHDL hardware compiler, in *Proceedings of the International Conference on Field Programmable Logic and Applications (FPL)* (2012), pp. 619–622
27. GAUT - High-Level Synthesis Tool From C to RTL, Accessed 21 Mar 2015. http://hls-labsticc.univ-ubs.fr/
28. G. Martin, G. Smith, High-level synthesis: past, present, and future. IEEE Des. Test Comput. **26**(4), 18–25 (2009)
29. W. Meeus, K. Van Beeck, T. Goedemé, J. Meel, D. Stroobandt, An overview of todays high-level synthesis tools. Des. Autom. Emb. Syst., pp. 1–21 (2012)
30. BDTI, An Independent Evaluation of the AutoESL AutoPilot High-Level Synthesis Tool (2010). http://www.bdti.com/Resources/BenchmarkResults/HLSTCP/AutoPilot. Accessed 10 Oct 2012
31. F. Winterstein, S. Bayliss, G. Constantinides, High-level synthesis of dynamic data structures: a case study using Vivado HLS, in *Proceedings of the International Conference on Field-Programmable Technology (ICFPT)* (2013), pp. 362–365

32. R. Nane, V.-M. Sima, C. Pilato, J. Choi, B. Fort, A. Canis, Y.T. Chen, H. Hsiao, S. Brown, F. Ferrandi, J. Anderson, K. Bertels, A survey and evaluation of FPGA high-level synthesis tools, *IEEE Transactions on Computer-Aided Design of Integrated Circuits and Systems*. http://janders.eecg.toronto.edu/pdfs/tcad_hls.pdf. Accessed 28 Feb 2016
33. S. Sarkar, S. Dabral, P. Tiwari, R. Mitra, Lessons and experiences with high-level synthesis. IEEE Des. Test Comput. **26**(4), 34–45 (2009)
34. P. O'Hearn, J. Reynolds, H. Yang, Local reasoning about programs that alter data structures, in *Computer Science Logic*, ed. by L. Fribourg, Lecture Notes Series, in Computer Science, vol. 2142, (Springer, Heidelberg, 2001), pp. 1–19
35. F.J. Winterstein, S.R. Bayliss, G.A. Constantinides, Separation logic for high-level synthesis. ACM Trans. Reconfigurable Technol. Syst. **9**(2), 10:1–10:23 (2015)
36. F. Winterstein, K. Fleming, H.-J. Yang, J. Wickerson, G. Constantinides, Custom-sized caches in application-specific memory hierarchies, in *Proceedings of the International Conference on Field Programmable Technology (ICFPT)* (2015), pp. 144–151
37. F. Winterstein, K. Fleming, H.-J. Yang, S. Bayliss, G. Constantinides, MATCHUP: memory abstractions for heap manipulating programs, in *Proceedings of the ACM/SIGDA International Symposium on Field-Programmable Gate Arrays (FPGA)* (2015), pp. 136–145
38. F. Winterstein, S. Bayliss, G.A. Constantinides, Separation logic-assisted code transformations for efficient high-level synthesis, in *Proceedings of the IEEE International Symposium on Field-Programmable Custom Computing Machines (FCCM)* (2014), pp. 1–8
39. F. Winterstein, S. Bayliss, G. Constantinides, FPGA-based K-means clustering using tree-based data structures, in *Proceedings International Conference on Field Programmable Logic and Applications (FPL)* (2013), pp. 1–6
40. Vivado-KMeans: Hand-Written HDL Code and C-Based HLS Designs for K-means Clustering Implementations on FPGAs. https://github.com/FelixWinterstein/Vivado-KMeans. Accessed 19 Dec 2015
41. K. Fleming, H.-J. Yang, M. Adler, J. Emer, The LEAP FPGA operating system," in *Proceedings of the International Symposium on Field Programmable Logic and Applications (FPL)* (2014), pp. 1–8
42. H.-J. Yang, K. Fleming, M. Adler, F. Winterstein, J. Emer, LMC: automatic resource-aware program-optimized memory partitioning, in *Proceedings of the ACM/SIGDA International Symposium on Field-Programmable Gate Arrays (FPGA)* (2016), pp. 128–137
43. H.-J. Yang, K. Fleming, M. Adler, F. Winterstein, J. Emer, Scavenger: automating the construction of application-optimized memory hierarchies, in *Proceedings of the International Conference on Field Programmable Logic and Applications (FPL)* (2015), pp. 1–8
44. The LEAP Run-time System - Rapid System Integration of Your HLS Kernels. http://www.fpl2015.org/?page=tutorials. Accessed 03 Mar 2016
45. LEAP-HLS: Rapid System Integration of High-level Synthesis Kernels Using the LEAP FPGA Framework. https://github.com/FelixWinterstein/LEAP-HLS. Accessed 31 Aug 2015
46. N. Ramanathan, J. Wickerson, F. Winterstein, G.A. Constantinides, A case for work-stealing on FPGAs with OpenCL atomics," in *Proceedings of the ACM/SIGDA International Symposium on Field-Programmable Gate Arrays (FPGA)* (2016), pp. 48–53
47. S.T. Fleming, D.B. Thomas, F. Winterstein, *FPGAs and Parallel Architectures for Aerospace Applications: Soft Errors and Fault-Tolerant Design*, pp. 75–90 (Springer International Publishing, Heidelberg, 2016). (ch. A Power-Aware Adaptive FDIR Framework Using Heterogeneous System-on-Chip Modules)
48. E.A. Pelaez, S. Bayliss, A. Smith, F. Winterstein, D.R. Ghica, D. Thomas, G.A. Constantinides, Compiling higher order functional programs to composable digital hardware, in *Proceedings of the IEEE International Symposium on Field-Programmable Custom Computing Machines (FCCM)* (2014), pp. 234–234

# Chapter 2
# High-Level Synthesis of Dynamic Data Structures

HLS promises significant shortening of the design cycle compared to a design entry at RTL. However, many HLS implementations require extensive code alterations to ensure synthesisability and to achieve latency, throughput and resource utilisation comparable to handwritten RTL designs. These are especially important for programs with 'irregular control flow' and 'complicated data dependencies'. In this chapter, we describe these terms in detail and elaborate on their implications for efficient HLS. To this end, we present a case study comparing the implementations of two algorithms for a compute-intensive machine learning application ($K$-means clustering). Algorithmically, both implementations solve the same problem, but they differ significantly in their computational properties: the first is a data flow-centric, 'regular' implementation with simple control flow, whereas the second is based on a recursive traversal of a pointer-linked tree data structure and uses dynamic memory allocation. The latter application thus exhibits highly 'irregular control flow' and 'complicated data dependencies'. Despite this irregularity, software implementations of this algorithm have been shown to be significantly faster than their data flow-centric counterparts because it effectively reduces the algorithmic complexity of the problem [1].

Our evaluation fits in the line of works that present designer's experiences with HLS tools. For example, a broad selection of 12 state-of-the-art HLS tools, academic and commercial, is evaluated by Meeus et al. [2]. Their overview, attesting Vivado HLS excellent test results, targets FPGA as well as ASIC flows and is based on a large set of criteria grouped into language support, ease of use, QoR and the capability of a rapid design space exploration. The goal is to perform a broad comparison across different tools mainly using a *Sobel* edge detector [3] as a test case. Sarkar et al. [4] present a more refined designer's experience with three HLS tools for ASICs using stream-based video processing applications. Their conclusion highlights the importance of fine-grained re-architecturing their test cases to optimise area and power consumption, and an evaluation by experienced users to obtain solid comparisons. BDTI present an explicit evaluation of AutoPilot (later renamed into Vivado HLS after the acquisition by Xilinx) [5]. Their evaluation uses video processing and stream-based wireless communications benchmarks, reporting QoR comparable with



F. Winterstein, *Separation Logic for High-level Synthesis*, Springer Thesis,
DOI 10.1007/978-3-319-53222-6_2

manual RTL implementations. The evaluations above share the commonality that the chosen benchmark cases are data flow-centric stream-based applications with simple control flow. A recent survey in [6] compares three academic tools and one commercial HLS tool using and four data-flow centric benchmarks in addition to the CHStone [7] benchmark suite, which covers a broader spectrum of applications. Heap-manipulating code, however, is not included. In contrast to the above evaluations, with our pointer-based benchmark, we aim to operate the HLS flow on test cases outside its 'comfort zone'.

The outcome of our case study is three-fold: Firstly, we can show that the performance result obtained for software implementations can be repeated with hand-optimised RTL implementations of both algorithms. This result is interesting in that irregular algorithms are often believed to be inefficient once mapped into hardware. Furthermore, it shows that the use of dynamic, pointer-linked data structures, which are central to the second algorithm, can result in very efficient FPGA applications if implemented well. Secondly, we repeat the case study with an HLS implementation using a state-of-the-art HLS tool and show that our previous result is reversed if the source code is not substantially altered prior to HLS. Thirdly, we analyse the efficiency with which the HLS tool maps specific program features into RTL and propose source-to-source transformations that improve the QoR of the irregular algorithm by a factor of eight in terms of latency, significantly narrowing the gap between HLS and hand-written RTL implementations. This chapter describes:

- An efficient RTL implementation of the irregular tree-based $K$-means clustering algorithm which preserves the algorithmic advantage over the conventional regular implementation. We show how the implementation can efficiently exploit the distributed memory architecture in FPGAs.
- A comparative case study using a data-flow centric clustering implementation and an implementation based on recursive traversal of a pointer-linked tree structure which incorporates data-dependent control flow. The case study comprises hand-written RTL and HLS implementations. Code transformations necessary to enable HLS of unsupported program features are highlighted.
- The use of on-chip dynamic memory allocation which allows us to allocate the average amount of memory required during runtime instead of statically pre-allocating the worst-case amount resulting in a 57× reduction of on-chip memory resources.
- An end-to-end QoR comparison between the automatically generated RTL code for both variants and both functionally equivalent, hand-written RTL implementations.
- An analysis of how efficiently specific program features are synthesised into RTL. We propose source-to-source transformations that improve QoR by a factor of eight in terms of latency.

The two algorithms for $K$-means clustering form the basis of our case study. Figure 2.1 shows our design flow. The initial C++ model is modified in order to include custom precision for operands of the basic arithmetic operations. From this model, we implement a hand-written RTL design written in VHDL (bottom branch, Sect. 2.3) and a C++-based HLS design (top branch). The HLS

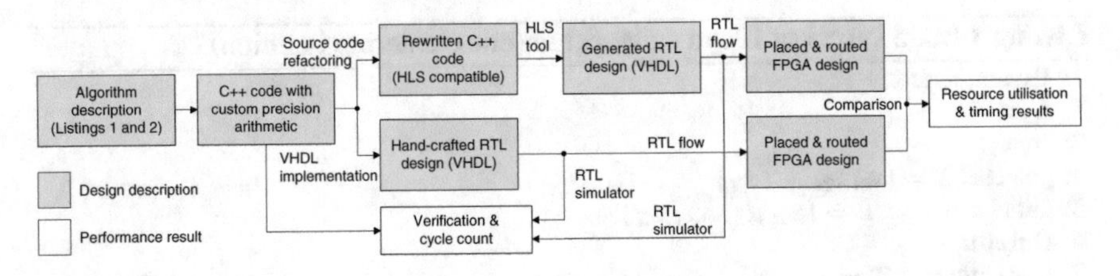


**Fig. 2.1** Design flow of the case study

implementation requires further code refactoring which we discuss in Sect. 2.4. The generated and hand-crafted RTL design entries are verified using standard RTL simulation tools. Finally, QoR is compared in terms of latency and resource usage taken from the placed and routed FPGA designs (Sect. 2.5). The evaluation flow in Fig. 2.1 is repeated for both clustering algorithms. The following section discusses both algorithms.

## 2.1 Background

The test cases we chose for this case study are two implementations of a clustering application, a technique for unsupervised partitioning of a data set commonly used in a wide range of applications, such as machine learning and data mining [8, 9], radar tracking [10], image colour or spectrum quantisation [11–14]. A popular technique for finding clusters in a data set is $K$-means clustering, which partitions the $D$-dimensional point set $X = \{x_j\}, j = 1, \dots, N$ into clusters $\{S_i\}, i = 1, \dots, K$, where $K$ is provided as a parameter. The goal is to find the optimal partitioning which minimises the total sum of squared Euclidean distances (squared-error distortion) given in (2.1) where $\mu_i$ is the geometric centre (centroid) of $S_i$.

$$J(\{S_i\}) = \sum_{i=1}^{K} \sum_{x_j \in S_i} \left\| x_j - \mu_i \right\|^2 \quad (2.1)$$

Finding optimum solutions to this problem is NP-hard [15]. A popular heuristic version uses an iterative refinement scheme. The standard algorithm begins by choosing $K$ initial centres $Z = \{\mu_1, \dots, \mu_K\}$ sampled randomly from the point set. The set $Z$ is iteratively refined until it no longer changes. On each iteration, it splits $X$ into $K$ partitions, according to which is the nearest mean of each partition. These means (geometrical centres) form the next generation of $Z$ ($Z'$). Using one algorithm for this problem, which we refer to as *Lloyd's algorithm*, $N \cdot K \cdot L$ distances in $D$-dimensional space are computed where $N$ is the number of data points and $L$, the number of required iterations. Listing 1 shows pseudo code of the main processing loop for one iteration of Lloyd's algorithm. Line 12 searches among $K$ candidate

**Listing 1** Main kernel of Lloyd's algorithm (one clustering iteration).

1: **Parameters:**
2: $N, K$
3: **Input**:
4: point set $X = \{x_1, x_2, \ldots, x_N\}$
5: initial centre set $Z = \{\mu_1, \mu_2, \ldots, \mu_K\}$
6: **Output**:
7: new centre set $Z' = \{\mu'_1, \mu'_2, \ldots, \mu'_K\}$
8: **Variables:**
9: centroid information $C = \{c_1, c_2, \ldots, c_K\}$

10: **function** LLOYDS
11: **for all** $x_j \in \{x_1, x_2, \ldots, x_N\}$ **do** ▷ iterate over all data points
12: $i \leftarrow \text{argmin}_{i',\mu_{i'} \in Z}(||x_j - \mu_{i'}||^2)$ ▷ find closest centre to $x_j$ among $K$ candidates
13: $c_i \leftarrow$ select $i$th element in $C$
14: $c_i.wgtCent \leftarrow c_i.wgtCent + x_j$
15: $c_i.count \leftarrow c_i.count + 1$
16: update $c_i$ in $C$
17: **end for**
18: **for all** $c_i \in C$ **do** ▷ update centre positions
19: $\mu'_i \leftarrow c_i.wgtCent/c_i.count;$
20: **end for**
21: **end function**

centres for the closest centre to a data point $x_i$. The index $i$ of this centre is used to update the correct entry in the centroid information table $C$ (Lines 13–16). $C$ contains $K$ vector sums of data points which we refer to as 'weighted centroids' ($wgtCent$). After all data points have been processed, the final output centre set $\{\mu'_1, \mu'_2, \ldots, \mu'_K\}$ is computed from the weighted centroids in $C$ (Lines 18–20).

In contrast to massively parallel hardware implementations, sophisticated software implementations have been proposed which gain speed-up from search space reductions. Kanungo et al. [1] present one possible implementation. Their *filtering algorithm* organises the data points in a multi-dimensional binary search tree, called a 'kd-tree', and finds nearest centres at each iteration using a tree traversal. To this end, the point set is recursively divided into two subsets. In each step, the axis-aligned bounding box of the subset is computed and subdivided. This leads to a (generally not perfectly balanced) binary kd-tree structure whose root node represents the bounding box of all data points and whose children nodes represent recursively refined, non-empty disjoint bounding boxes. Each tree node stores the bounding box ($bndBox$) information as well as the number ($count$) and the vector sum of its associated points (the weighted centroid, $wgtCent$) which is used to update the cluster centres when each iteration completes. The weighted centroid of leaf nodes is the data point itself.

Listing 2 shows a simplified version of the recursive kernel function of the filtering algorithm for one iteration. During clustering, the tree is traversed starting from the root node. The set of input centres in Lloyd's algorithm is replaced by sets of candidates for the closest centre to a subset of data points. The algorithm propagates multiple candidate sets down the tree. These are of variable size and are created and disposed at run-time. At each non-terminal visited tree node, the closest candidate centre to the mid point ($midPoint$) of the bounding box is found. Some of the remaining candidates are pruned if no part of the bounding box is closer to them than the closest centre (Line 22). The pruning greatly reduces the number of computed distances since the average number of 'close' cluster-centre candidates is significantly smaller than $K$. Additionally, entire sub-trees can be pruned if only one candidate remains. As the point set does not change during clustering, the kd-tree needs to be built up only once and the additional overhead is amortised over all iterations. In fact, our profiling results show that, on average, the tree construction demands less than 2% of the total computation required. Therefore, we perform the pre-processing in software and the FPGA accelerator discussed in the following focuses only on the tree traversal phase.

In light of this case study, we identify the most important features of both applications. Because the min-search in Listing 1 (Line 12) is implemented as a `for`-loop over $K$ centres, the main kernel of Lloyd's algorithm consists of two nested `for`-loops with constant bounds. The simple control flow and inherent parallelism at the granularity of distance computations makes the computationally expensive algorithm suitable for hardware implementations so as to accelerate $K$-means clustering for real-time implementations if $N$ and $K$ are large. Computational parts of the filtering algorithm in Listing 2 are the closest centre searches (Lines 14, 20) and the candidate pruning (Line 22, containing two distance calculations), and the centroid buffer update. The loops in the min-searches and candidate pruning have variable bounds $2 \leq k \leq K$. The implementation uses dynamic memory allocation (Line 21) and de-allocation (Lines 32, 36) enclosed in data-dependent conditionals. Memory space is freed upon backward traversal, i.e. after an allocated centre set has been read twice. The implementation uses recursive function calls (beyond tail recursion) which requires the presence of a *stack*. The stack is implicitly handled in the software program, but it needs to be explicitly implemented in an FPGA application. The data passed between recursive instances are the tree node $u$ and the set of candidate centre set $Z$.

Previous hardware implementations of Lloyd's algorithm are proposed in [14, 16–19]. Pioneering work by Leeser et al. [16] implemented FPGA-clustering for the analysis of hyperspectral images. Their approach trades clustering quality for hardware resource consumption by replacing the Euclidean distance norm with multiplier-less Manhattan and Max metrics. This trade-off is extended to bit width truncations on the input data by Estlick et al. [14] who report a speed-up of up to 200× over the software implementation. More recent work in [17] builds on the same framework and extends it by incorporating a hybrid fixed- and floating-point arithmetic architecture. These approaches aim to gain acceleration from an increased amount of parallel hardware resources for distance computations and nearest centre search.

**Listing 2** Main kernel of the filtering algorithm (one clustering iteration) [1].

1: **Parameters**:
2: $N, K$
3: **Input**:
4: kd-tree
5: initial centre set $\{\mu_1, \mu_2, \ldots, \mu_K\}$
6: **Output**:
7: new centre set $Z' = \{\mu'_1, \mu'_2, \ldots, \mu'_K\}$
8: **Variables**:
9: node in the kd-tree $u$
10: multiple sets of candidates for the closest centre to a point cloud ($Z$)
11: centroid information $C = \{c_1, c_2, \ldots, c_K\}$

12: **function** FILTER($u$, $Z$)
13: **if** $u$ is leaf **then**
14: $i^* \leftarrow \text{argmin}_{i', \mu_{i'} \in Z}(||u.wgtCent - \mu_{i'}||^2)$ ▷ find closest centre to $u.wgtCent$
15: $c_{i^*} \leftarrow$ select $i^*$-th element in $C$
16: $c_{i^*}.wgtCent \leftarrow c_{i^*}.wgtCent + u.wgtCent$
17: $c_{i^*}.count \leftarrow c_{i^*}.count + 1$
18: update $c_{i^*}$ in $C$
19: **else**
20: $i^* \leftarrow \text{argmin}_{i', \mu_{i'} \in Z}(||u.midPoint - \mu_{i'}||^2)$ ▷ find closest centre to $u.midPoint$
21: $Z_{new} \leftarrow$ **new** centre set ▷ allocate new centre set (empty)
22: **for all** $\mu_j \in Z$ **do** ▷ prune candidate centres
23: **if** pruningTest($i^*$, $\mu_j$, $u.bndBox$) is false **then**
24: $Z_{new} \leftarrow Z_{new} \cup \{\mu_j\}$; ▷ insert surviving candidates into $Z_{new}$
25: **end if**
26: **end for**
27: **if** $|Z_{new}| = 1$ **then**
28: $c_{i^*} \leftarrow$ select $i^*$-th element in $C$
29: $c_{i^*}.wgtCent \leftarrow c_{i^*}.wgtCent + u.wgtCent$
30: $c_{i^*}.count \leftarrow c_{i^*}.count + u.count$
31: update $c_{i^*}$ in $C$
32: **delete** $Z_{new}$ ▷ immediately delete allocated $Z_{new}$
33: **else** ▷ recurse on children
34: FILTER($u.left$ , $Z_{new}$);
35: FILTER($u.right$, $Z_{new}$);
36: **delete** $Z_{new}$ ▷ delete allocated $Z_{new}$ on the way back
37: **end if**
38: **end if**
39: **end function**
40: **for all** $c_i \in C$ **do** ▷ update centre positions
41: $\mu'_i \leftarrow c_i.wgtCent/c_i.count$;
42: **end for**

Contrary to these works, the first contribution in this thesis chapter is an efficient implementation of the filtering algorithm, which gains acceleration largely from search space pruning. Chen et al. [20] present a VLSI implementations for $K$-means clustering which is notable in that it, in line with our approach, recursively splits the data point set into two subspaces using conventional 2-means clustering. Logically, this creates a binary tree which is traversed in a breadth-first fashion and results in computational complexity proportional to $\log_2 K$. This approach, however, does not allow any pruning of candidate centres. Saegusa et al. [12] present a simplified kd-tree-based implementation for $K$-means image clustering. The data structure stores the best candidate centre (or generally a few 'best' candidates) at its leaf nodes and is looked up for each data point. The tree is built independently of the data points, i.e. the pixel space is subdivided into regular partitions which leads to 'empty' pixels being recursively processed. Other disadvantages are that the tree needs to be rebuilt at the beginning of each iteration and that the centre lists are not pruned during tree traversal in the build phase, which are essential features of the filtering algorithm.

## 2.2 Analysis of the Filtering Algorithm

We analyse several properties of the filtering algorithm that provide insight into the advantage over Lloyd's algorithm. To this end, we profile a software implementation of the algorithm. The input data sets that we use throughout this chapter are point sets of $N = 16{,}384$ three-dimensional real-valued samples. The data points are distributed among 128 centres following a normal distribution with varying standard deviation $\sigma$, whereas the centre coordinates are uniformly distributed over the interval $[-1, 1]$. Finally, the data points are converted to 16bit fixed-point numbers. We choose $K = 128$ initial centres sampled randomly from the data set and run the algorithm either until convergence of the objective function or until a maximum of 30 iterations are reached. In addition to synthetic input data, we include a working set with $N = 16,384$ randomly sampled pixels from the well-known Lena benchmark image and quantise the colour space into $K = 128$ clusters. Note that the clustering output is exactly the same for both the implementation of Lloyd's and the filtering algorithm.

The filtering algorithm can be divided into two phases: building the tree from the point set (pre-processing), and the repeated tree traversal and centre update (clustering phase). In order to obtain information about the computational complexity of both parts, we profile the software implementation of the algorithm using synthetic input data. Here, we chose the number of Euclidean distance computations performed as our metric for computational complexity. Since the tree creation phase does not compute any distances but performs mainly dot product computations and comparisons, we introduce distance computation equivalents (DCEs) to obtain a unified metric for both parts which combines several operations which are computationally equivalent. Table 2.1 shows the profiling results of the computational complexity of the filtering algorithm broken down into clustering and pre-processing phases for different working sets. The parameter $\sigma$ is varied such that the synthetic input data ranges from well-distinguished clusters ($\sigma = 0.05$) to a nearly unclustered point set ($\sigma = 0.35$). For all cases, the number of DCEs performed during tree creation is only a fraction of the total number of DCEs (2% geometric mean). Because of the small contribution of

**Table 2.1** Computational complexity of the filtering algorithm broken down into clustering and pre-processing phases

| Input data $N = 16,384$, $K = 128$ | DCEs in clustering | DCEs in pre-processing | Contribution of pre-processing (%) |
|---|---|---|---|
| Synthetic $\sigma = 0.05$ | 1,09,207 | 4963 | 4.3 |
| Synthetic $\sigma = 0.10$ | 1,56,464 | 4712 | 2.9 |
| Synthetic $\sigma = 0.15$ | 2,12,670 | 4574 | 2.1 |
| Synthetic $\sigma = 0.20$ | 2,59,146 | 4494 | 1.7 |
| Synthetic $\sigma = 0.25$ | 2,94,173 | 4423 | 1.5 |
| Synthetic $\sigma = 0.30$ | 3,21,841 | 4432 | 1.4 |
| Synthetic $\sigma = 0.35$ | 3,39,831 | 4424 | 1.3 |
| Lena benchmark (subset) | 2,24,418 | 4923 | 2.1 |

the pre-processing, we perform this part in software and the FPGA implementation described in the following section focuses on the tree traversal phase only.

We also evaluate the search space pruning. The major complexity reduction is due to the fact that the repeated searches for the closest centre need to consider significantly fewer centres than Lloyd's algorithm for which this number is always $K$. Figure 2.2 (left) shows the frequency of candidate centre set sizes averaged over all synthetic cases above. During tree processing, most sets contain only 2 or 3 centres and the average centre set size is 4.36 (3.78 for the Lena image benchmark), which shows the effectiveness of the search space pruning. We quantify the overall search complexity of the filtering algorithm in terms of the aggregate number of *node-centre pairs*, i.e. the cumulative number of candidate centres processed at the visited tree nodes per clustering iteration. This number is sensitive to the input data.

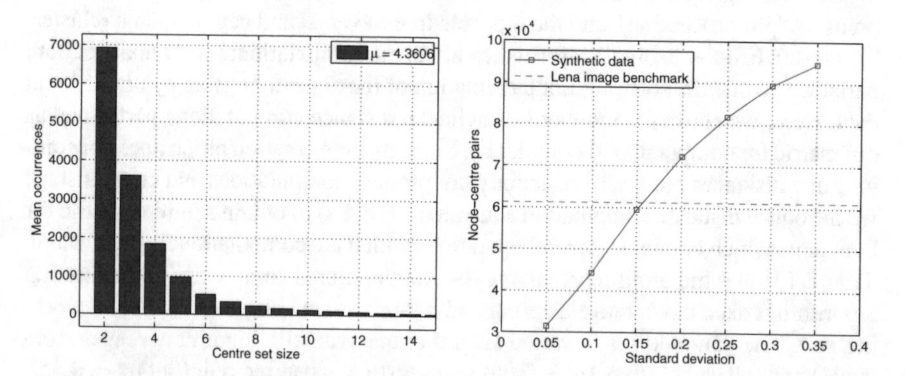


**Fig. 2.2** *Left* Frequency of candidate centre set sizes for synthetic input data. *Right* Computational complexity of the filtering algorithm in terms of node-centre pairs (Lloyd's algorithm has a constant complexity of $209.7 \times 10^4$ point-centre pairs for this data set)

Figure 2.2 (right) shows the number of node-centre pairs over different values of $\sigma$ in the synthetic data sets. The complexity ranges from 31,399 to 94,590. We also include the Lena benchmark with 61,230 node-centre pairs for a comparison with real-world data. For Loyd's algorithm, an equivalent metric of data point-centre pairs can be defined which is $N \cdot K$ =20,97,152 for all input sets in Fig. 2.2. Even for unfavourable input data ($\sigma = 0.35$), the filtering algorithm thus achieves a 22× reduction of search complexity. In a sequential software implementation [1], this reduction translates directly into a run-time advantage of the filtering algorithm. The next sections investigate if, how, and to what extent this result can be reproduced in hardware implementations.

## 2.3 RTL Implementations

This section describes efficient hand-crafted FPGA implementations of Lloyd's and Kanungo's filtering algorithm implementations, which will be compared in Sect. 2.5.1. Both RTL implementations are fully pipelined designs and their computational parts mainly consist of the same basic elements, Euclidean distance and dot product computations, but their control structures and memory architectures are substantially different. We made the source code of the RTL implementations discussed below available in an open source repository.[1] The following description motivates later discussion of how we direct the HLS flow to produce competitive designs from a C description. Specific features discussed here and implemented later in the HLS flow (Sect. 2.4) will disclose particular limitations.

### 2.3.1 Lloyd's Algorithm

The implementation consecutively fetches data points from memory, computes the Euclidean distance to each centre $\mu_i$, $1 \leq i \leq K$, and selects the closest centre before fetching the next data point. The distance computation is fully parallelised for a parametric data point dimensionality $D$. Parallelism is further increased by performing $P$ distance computations concurrently which reduces the number of sequential steps per iteration from $N \cdot K$ to $(N \cdot K)/P$. A centroid buffer stores the centroid information $C$ and maintains the intermediate results during one iteration which are continuously updated. The accumulated weighted centroids ($wgtCent$) are then divided by the $count$ value at each index to obtain the centre positions for the next iteration. The data set memory and centroid buffer are implemented as on-chip block random access memory (BRAM) and distributed look-up table (LUT) RAM, respectively. The position update uses a pipelined divider core.

[1]https://github.com/FelixWinterstein/Vivado-KMeans [21].

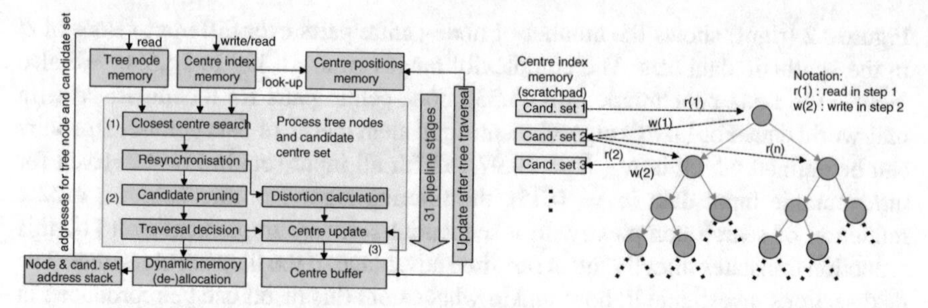


**Fig. 2.3** *Left* FPGA implementation of the filtering algorithm. *Right* Read-write accesses to the scratchpad memory for centre sets during tree traversal

## 2.3.2 Filtering Algorithm

Figure 2.3 (left) shows a high-level block diagram of our RTL design of the filtering algorithm. Our RTL implementation contains three computational kernels: (1) The closest centre search computes Euclidean distances to either the mid point of a bounding box or the tree node's weighted centroid, followed by a min-search. (2) The pruning kernel performs two slightly modified distance computations to decide whether any part of the bounding box crosses the hyperplane bisecting the line between two centres. A more detailed description of the pruning algorithm is given in [1]. Those centres $\mu_i$ for which the pruning test returns false are flagged and no longer considered by subsequent processing units. (3) The centroid buffer is updated and used in the same way as for Lloyd's algorithm. All three sub-kernels are integrated in a pipelined, stream-based processing core. This core has a hardware latency of 31 clock cycles and can accept a node-centre pair on every other clock cycle. Thus, if fully utilised, the pipeline is usually filled with several tree nodes and their associated candidate centre sets.

The heart of the filtering algorithm is the traversal of the kd-tree which is implemented using the recursive calls shown in Listing 2. Our implementation controls this tree traversal using a stack which contains pointers to a tree node and to its associated set of candidate centres as well as the current set size. After fetching the pointers from stack, the data referenced by them is processed. At the output of the pipeline, we obtain a new traversal decision which is based on whether we have not yet reached a leaf node and whether there is more than one centre in the pruned candidate set left. If so, new pointers (left and right child and a new centre set) and the new set size are pushed onto the stack. Otherwise, nothing is pushed onto the stack. In the latter case, a pointer to a non-visited node further up in the tree will be fetched for processing in the next cycle. This process is repeated until the stack and pipeline are empty which terminates the tree traversal. Because all memories (tree nodes, centre indices, centre positions, centroid buffer, and stacks) are mapped to physically disjoint memories, all accesses can be made simultaneously in each clock cycle.

### Pipelining and Parallelisation

The profiling results in Sect. 2.2 show that a candidate set (associated with a tree node and processed item by item) has an average size of 4.36 centres in the scenarios considered here, which is smaller than the pipeline depth of 31 stages. In order to hide pipeline latency, we need to overlap the processing of multiple node-centre set pairs in the pipeline, which is possible in the absence of feedback dependencies. Figure 2.3 (right) illustrates the read and write accesses. Memory accesses are indicated by dashed lines, pointer links are drawn as solid lines. The diagram shows that a read-write data dependency exists only between centre sets whose associated tree nodes have a direct parent-child relation. In fact, all pointers residing on the stack point to data structures that has already been written to and hence can be processed independently. The scheduler in the stack management fetches new pointers as described above as soon as the pipeline is ready to accept new data. Independent centre sets are read and written simultaneously using dual-port memory. For parallelism beyond pipelining the processing units are duplicated. To process independent subsets of such pairs, we split the tree into $P$ disjoint sub-trees and distribute them across several computational units for parallel processing. We note that for both pipelining and parallelisation, we exploit knowledge about dependencies carried by data structures accessed through pointers.

### Dynamic Memory Allocation

The centre index memory (Fig. 2.3, left) serves as a *scratchpad* memory for storing centre sets and retaining them for later usage during the tree traversal. A new set is written when child nodes are pushed onto the stack and must be retained until both left and right child nodes have been processed. The memory space then can be freed and reused. The duration for which a centre set must be retained in memory depends on the shape of the (generally unbalanced) tree. The results in Sect. 2.2 are obtained under the assumption that the application can allocate as much scratchpad memory as needed. However, the requested amount may exceed the available on-chip memory resources. The worst-case number of candidate sets is $N - 1$ which is required in the case of a degenerate kd-tree where every internal node's right child is a leaf and its left child is another internal node. If we consider an FPGA application supporting $N_{max} = 16{,}384$ data points and a maximum of $K_{max} = 256$ centres, we require $(N_{max} - 1) \cdot K_{max} \cdot \log_2 K_{max} \approx 33.6$ Mbits worst-case memory space which consumes 912 on-chip 36 k-BRAM resources ($\sim$89% in a medium-size Virtex 7 FPGA) and does not leave enough resources for the other memories in the implementation. However, in the average case, the tree is unlikely to be degenerate as described above and therefore the lifetime of a centre set is much shorter and the instantaneous memory requirement is significantly lower.

As a result of this resource advantage, we implement a memory management unit which dynamically allocates space and frees it once the candidate set has been read for the second time, rather than a static allocation. The implementation of the fixed-size allocator uses a *free-list* that keeps track of occupied memory space. In our implementation, the scratchpad memory and free-list are sized to accommodate an

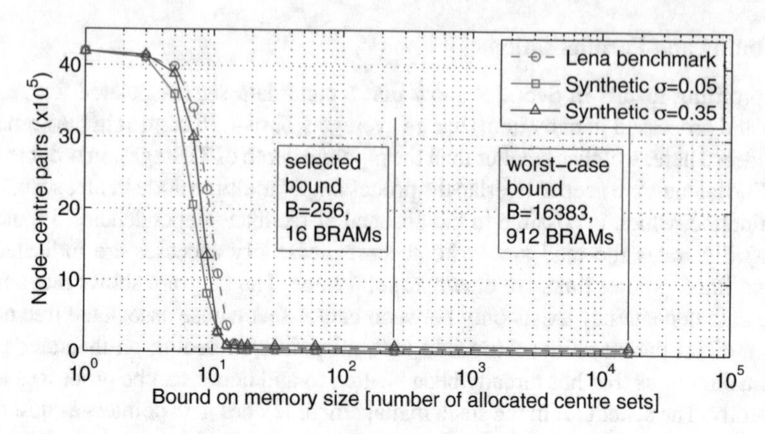


**Fig. 2.4** Trade-off between heap size and run-time of the filtering algorithm (profiling)

'average-case' number of centre-candidate sets. Our approach is to limit the memory to a size of $B \ll N - 1$ sets. When inadequate memory is available to service an allocation request, the algorithm allows us to abandon the pruning approach and instead consider all candidate centres. This modification does not compromise the functionality of the algorithm, but it increases its run-time (the number of node-centre interactions). Figure 2.4 shows the result of profiling the software implementation clustering N = 16,384 pixels (RGB vectors) sampled from the Lena image benchmark and the two extreme cases for synthetic data in Table 2.1. If we allow the algorithm to allocate memory for only a single centre, the search complexity degrades to the worst case of $(2 \cdot N - 1) \cdot K$ node-centre pairs to be examined. The search complexity, however, greatly decreases for $B > 10$ in all test scenarios. We select a bound of $B = 256$ centre sets (16 36k-BRAMs) which practically causes no run-time degradation in the scenarios considered in this case study.

The next section describes the re-implementations of both algorithms using a C-based HLS tool, which finally allows us to compare the FPGA resource usage and speed of all four designs.

## 2.4 HLS Implementations

We choose Vivado HLS for this case study as an exemplary state-of-the-art tool which shares many similarities with other modern C-to-FPGA flows such as LegUp [22], ROCCC [23], Dwarv [24] and GAUT [25]. RTL generation is guided by synthesis directives which are manually invoked and configured. Exploring design options and optimisations using directives ideally does not require the source code to be altered. The most important directives we use to control the RTL generation are loop pipelining and loop unrolling directives. Loop pipelining overlaps loop iterations in the pipeline. The interval between the start of two iterations is given by the initiation interval (II). Loop unrolling is used to force parallel instantiations of the loop body. In order to remove the bottleneck of an insufficient number of memory ports in a parallelised application, on-chip memories can be split into multiple banks using an *array partitioning* directive. As for LegUp, ROCCC, Dwarv and GAUT, the C-based input is restricted to a synthesisable subset. Vivado HLS allows pointers to be used as references to statically allocated arrays. However, it does not synthesise dynamic memory allocation (`new`, `delete`) and heap memory. In this thesis, we refer to pointer variables which obtain their value from a call to the `new` function as *heap-directed* pointers. Other disallowed features are system calls, arbitrary pointer casting and arbitrary recursive functions.

Our goal is to bring the generated RTL designs produced by the HLS flow as close as possible to the highly optimised manual RTL designs in the previous section. We distinguish between optimisations using synthesis directives and manual source code modifications.

## 2.4.1 Lloyd's Algorithm

The C code for Lloyd's algorithm corresponding to Listing 1 is directly synthesisable and does not contain any unsupported language features. We unroll all `for`-loops over the three dimensions of the input data points which results in a parallel implementation of the distance computation $||x_j - \mu_{i'}||^2$. Most of the computation is contained within the inner `for`-loop which implements the min-search in Line 12 (bound $K$). Pipelining this loop (II = 1) leads to performance comparable to hand-coded RTL. For acceleration beyond pipelining, we control the degree of parallelism just as in the case of the manual RTL design by partially unrolling the outer loop to degree $P$ (replicating pipelines). In order to match the parallelism of computational units and memory ports, we partition the centre positions and centroid buffer arrays into $P$ banks using the array partitioning directive. Overall, using synthesis directives and a minor source code modification to ensure correct indexing of the parallel instances of the centroid buffer, we are able to produce an RTL design which is architecturally similar to its hand-written counterpart.

## 2.4.2 Filtering Algorithm

The synthesisability of the main kernel as in Listing 2 requires the removal of the recursive function calls and the calls to `new` (Line 21) and `delete` (Lines 32, 36), and code transformations to improve QoR of the synthesis of the pointer-linked data structures and the circuits operating on these.

**Listing 3** Iterative replacement for the recursive kernel in Listing 2.

1: push to stack ($root$, $\{\mu_1, \mu_2, \ldots, \mu_K\}$, $true$);
2: **while** $stack$ not empty **do**
3: $u, Z, d \leftarrow$ fetch from head of stack
4: **if** ($d$ is `true`) **then**
5: **delete** $Z$
6: **end if**
7: $Z_{new} \leftarrow$ **new** centre set
8: … ▷ original body in Listing 2 (contains two variable-bound sub-loops)
9: **if** ($u$ is not a leaf) **and** ($|Z_{new}| > 1$) **then**
10: push to stack ($u.right$, $Z_{new}$, `true`)
11: push to stack ($u.left$, $Z_{new}$, `false`)
12: **else**
13: **delete** $Z_{new}$
14: … ▷ update centroid buffer
15: **end if**
16: **end while**

#### Recursive Tree Traversal

Recursion is replaced by a `while`-loop and a stack data structure. As in the RTL implementation, our C-based HLS design now contains three heap-allocated data structures: the pointer-linked kd-tree, the pool of centre sets and the stack. The program accesses these data structures through pointers. The stack contains the pointers to a heap-allocated tree node $u$ and a set of candidate centres $Z$ (and its size), as well as a flag $d$ indicating that the centre set can be de-allocated. Listing 3 shows the rewritten code that avoids recursion.

#### Dynamic Memory Allocation

We replace the basic C++ routines for dynamic memory allocation to ensure synthesisability by off-the-shelf HLS tools. Occurrences of `new` and `delete` statements are replaced by calls to custom allocator functions that we provide in an additional header file. The implementation of the fixed-size allocator is in Line with Sect. 2.3.2. Heap memory is replaced by arrays that are mapped to on-chip memory. We translate pointer dereferencing into array indexing and instantiate an array for each data structure type. We choose the same heap sizes as in the RTL implementation. The memory for centre sets is limited to the same bound $B$ as selected in Fig. 2.4. We implement the same fall-back solution when inadequate memory is available to service an allocation request as described in Sect. 2.3.2.

#### Parallelisation

As in the manual RTL design, we split the tree structure into $P$ independent subtrees to parallelise the application by instantiating $P$ parallel processing kernels. Heap memories for tree nodes and centre set memory are by default monolithic memory spaces which need to be divided into $P$ disjoint regions (sub-trees, and segments for private centre sets). The access through (dynamically allocated) pointers, however, hides this disjointness information, which renders the array partitioning directive

**Listing 4** Loop distribution to enable pipelining.

```
1: while stack not empty do
2:     while (stack not empty) and (queue not full) do
3:         u, Z, d ← fetch from head of stack
4:         enqueue (u, Z, d) in queue                    ▷ newly introduced queue
5:     end while
6:     while queue not empty do
7:         u, Z, d ← dequeue from queue
8:         ...                       ▷ remaining loop body in Listing 3 (Lines 4–15)
9:     end while
10: end while
```

ineffective and does not lead to parallel execution. In fact, applying automatic partitioning through HLS directives even leads to a degradation in latency as we show in the performance comparison in Sect. 2.5. Instead, we manually partition the tree node memory and privatise heap space for centre sets for each instance. This ensures that the scheduler of the HLS tool recognises the parallelisation opportunity. Automating this step requires a program analysis capable of identifying disjoint regions (in terms of access patterns) in the monolithic heap memory space.

### Inter-Iteration Dependencies and Pipelining

Apart from replication, acceleration of the manual RTL design is obtained from pipelining the tree traversal. This corresponds to pipelining the loop nest in Listing 3 which must take two (potential) inter-iteration dependencies into account. The first occurs between fetching pointers to data from the stack and pushing new pointers onto the stack, which hinders pipelining. However, because there are two `push` statements and one `fetch` statement, the items stored on the stack (pointers $u$ and $Z, d$) accumulate if the condition in Line 9 holds in several iterations. Once there are multiple pointers on the stack, these do not cause any read-write dependencies between iterations and hence can be overlapped in pipelined execution. Listing 4 shows a transformation of the loop in Listing 3 to implement this schedule. The transformation distributes the execution of the original loop body over two (pipelineable) inner loops which exchange data via a newly inserted queue. The second inner loop ensures that multiple items on stack will be immediately scheduled for processing. However, this loop still contains sub-loops with variable bounds which prevents the tool from pipelining it. An additional manual loop nest flattening transformation is required to enable pipelining the loop with II = 1. Because of the variable bounds of the inner loops, this loop nest is not a perfectly or semi-perfectly nested loop, which prevents the application of Vivados loop flattening directive. Without loop flattening, only the inner loops can be pipelined, which would result in less speed-up compared to the manually flattened loop.

The other (potential) inter-iteration dependency is due to the pointer references to $Z$ and $Z_{new}$ in Listing 3. This is a false dependency because, after the loop transformation, the pointers to $Z$ and $Z_{new}$ never alias across iterations. Inserting a 'dependence false' directive makes Vivado HLS aware of the non-existence of this dependency.

Enabling automatic pipelining for pointer-based programs thus crucially depends on an automated analysis capturing the semantics of `new` and `delete` and reasoning about such 'pointer-carried' dependencies which we will explore in Chap. 4.

## 2.5 Performance Comparison

We evaluate the four implementations (RTL and HLS designs for both algorithms) based on their execution time (latency) and resource consumption. For a latency comparison, we ran simulations on the synthetic data described in Sect. 2.2 for different values of $\sigma$. All hardware implementations produce the same clustering result as a software implementation that we implemented for validation. The algorithms ran until convergence or until 30 iterations were reached. All latency results below are per clustering iteration (average). This section begins with a comparison of the two RTL implementations. The latter part of the section then shows how close our manually optimised HLS designs can get to these results.

### *2.5.1 RTL Designs*

Figure 2.5 shows the average number of clock cycles per iteration of the FPGA-based filtering algorithm (left) as well as the average speed-up over the FPGA implementation of Lloyd's algorithm (right). We synthesise both RTL implementation of the filtering algorithm and Lloyd's algorithm for a Xilinx Virtex 7 FPGA (7vx485tffg-2) for varying degrees of parallelism. We use Xilinx Vivado 2014.4 for netlist

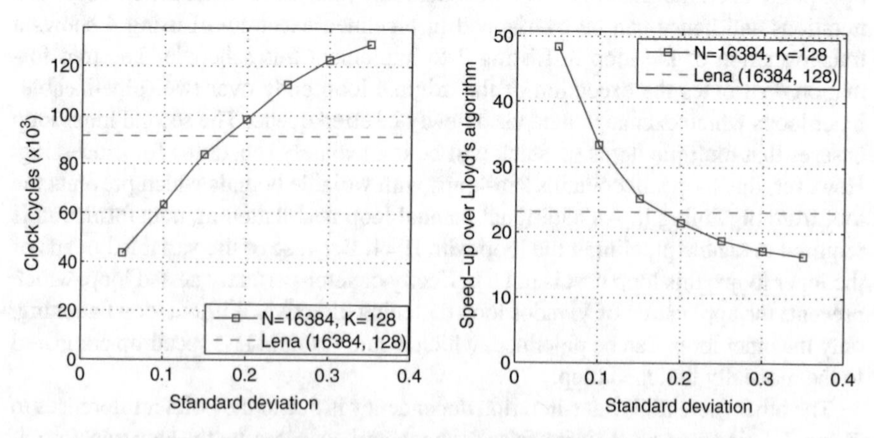


**Fig. 2.5** *Left* Average cycle count per iteration for the manual RTL implementation of the filtering algorithm ($P = 1$). *Right* Speed-up over an RTL implementation of Lloyd's algorithm ($P = 1$ in both cases)

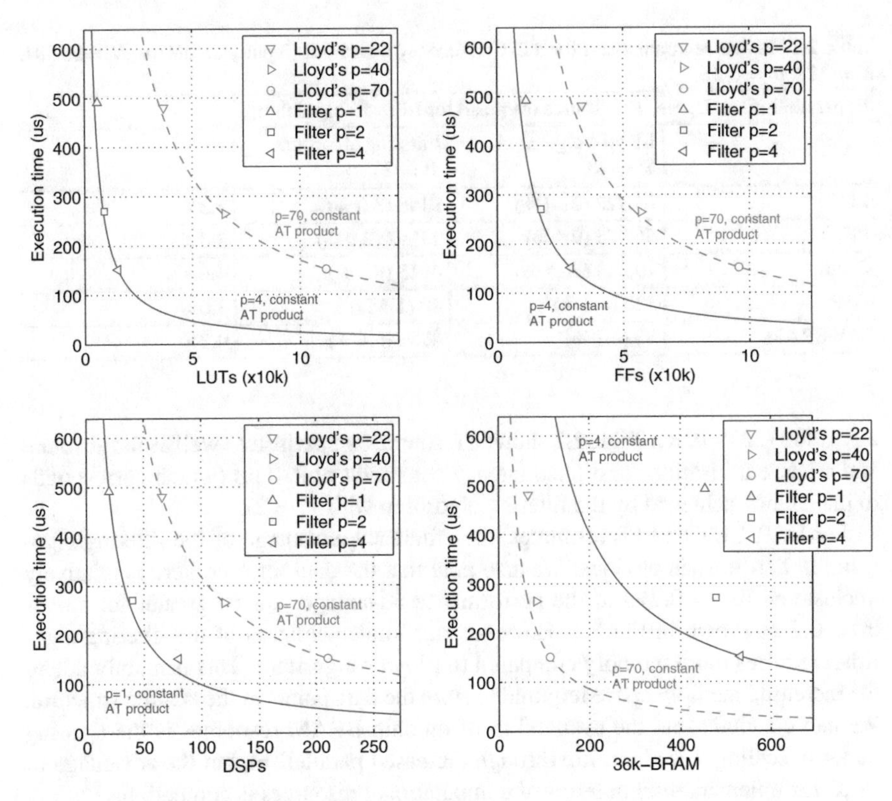


**Fig. 2.6** Mean execution time per iteration over FPGA resources for $N = 16384$, $K = 128$, $\sigma = 0.2$ (Xilinx Virtex7 7vx485tffg-2)

synthesis, placement and routing. We report the FPGA resource consumption for the different design points in terms of look-up tables (LUTs), flip-flops (FFs), FPGA slices (containing four LUTs and eight FFs), digital signal processing slices (DSPs) and 36 k-BRAM resources. All designs are synthesised for 200 MHz target clock frequency and all results are taken from fully placed and routed designs meeting the timing constraint. For the resource comparison of both implementations, we select the performance point in Fig. 2.5 with $\sigma = 0.2$, which lies amid the range of execution times and is close to the performance measured for the Lena benchmark. The degree of parallelism we choose in both implementations is given by the target latency which is expressed as average execution time per iteration. Figure 2.6 shows the area-time (AT) diagram, i.e. the amount of FPGA resources required to meet a target throughput. For ease of comparison of the two algorithms, we draw an area-time frontier with a constant AT product through the design points with the smallest AT product for each algorithm (solid blue and dashed red line; note that only the intersections of these lines with the data points are feasible designs). The inherent run-time advantage of the filtering algorithm needs to be countered by significantly increased parallelism of computational units in the implementation of Lloyd's

**Table 2.2** Resource comparison for a 270 μs-latency constraint (input parameters: $N$ =16,384, $K = 128$, $\sigma = 0.2$)

$P$: parallelisation degree, $R$: resource overhead for LLoyd's algorithm

| | Lloyd's algorithm $P = 40$ | Filtering algorithm $P = 2$ | $R$ |
|---|---|---|---|
| LUT | 64,922 (21.4 %) | 9148 (3.0 %) | 7.3× |
| FF | 56,975 (9.4 %) | 17397 (2.9 %) | 3.3× |
| Slices | 19,843 (26.1 %) | 4915 (6.5 %) | 4.0× |
| DSP | 120 (4.3 %) | 40 (1.4 %) | 3.0× |
| 36 k-BRAM | 83 (8.1 %) | 478 (46.4 %) | 0.2× |

algorithm (22×-70×). Table 2.2 shows a resource comparison as well as the absolute and relative utilisation for a fixed latency constraint of 270 μs (which corresponds to the latency achieved by the filtering algorithm with $P = 2$).

For DSP, LUT and FF resources, the efficiency advantage of the filtering algorithm in hardware is obvious. We also note that the data set used here is relatively unclustered ($\sigma = 0.2$) and the performance advantage will be greater for values $\sigma < 0.2$ as shown in Fig. 2.5. However, our implementation of the filtering algorithm requires more memory compared to Lloyd's algorithm. This is mainly due to the increased memory space required to store the data points in the kd-tree structure. We can conclude that the availability of on-chip BRAM resources is the limiting factor in scaling this algorithm through increased parallelism, but the advantage of its RTL implementation in terms of computational resources is compelling.

### 2.5.2 HLS Designs

We compare the performance of both HLS to both RTL designs based on different metrics: clock cycles count per iteration (through RTL simulations), execution time per iteration (includes the clock period), resource usage and AT product (in logic slices × ms). We implement the HLS designs with Xilinx Vivado HLS 2014.4. As in the previous section, all designs are synthesised for a 200 MHz target clock rate and all results are taken from fully placed and routed designs (not all designs meet the timing constraint in which case we report the best achievable clock period). The input data set to all implementations is the same data set as used above ($\sigma = 0.2$). In order to account for the inherent runtime advantage of the filtering algorithm due to search space pruning and to compare all four designs on a common basis, we increase the parallelisation degree for the final implementations of Lloyd's algorithm to $P = 40$, which equalises the cycle count of the hand-written RTL designs.

Table 2.3 shows the performance comparison based on the metrics above. The resource consumption of both HLS designs compared to their RTL counterparts is remarkably similar. The utilisation of flip flops is notable in that it is substantially

**Table 2.3** Performance comparison using the hand-written RTL designs as reference

Architecture: $N_{max} = 32,768$, $K_{max} = 256$, $B = 256$; input data (synthetic): $N = 16,384$, $K = 128$, $\sigma = 0.2$

| | Lloyd's algorithm | | Filtering algorithm | | | |
|---|---|---|---|---|---|---|
| | RTL (reference) | HLS | RTL (reference) | HLS (directives only) | HLS (manua partitioning) | HLS (manual loop transformation) |
| $P$ | 40 | 40 | 2 | 2 | 2 | 2 |
| Slices | 19,843 | 22,711 (×1.1) | 6950 | 5263 (×0.8) | 5161 (×0.7) | 6540 (×0.9) |
| LUT | 64,922 | 68,484 (×1.1) | 10,418 | 12,865 (×1.2) | 12,717 (×1.2) | 15046 (×1.4) |
| FF | 56,975 | 47,895 (×0.8) | 19,008 | 11,517 (×0.6) | 11,293 (×0.6) | 13612 (×0.7) |
| DSP | 120 | 120 (×1.0) | 40 | 36 (×0.9) | 36 (×0.9) | 36 (×0.9) |
| 36k-BRAM | 83 | 75 (×0.9) | 448 | 506 (×1.1) | 506 (×1.1) | 507 (×1.1) |
| Clock period | 5.0 ns | 8.4 ns (×1.7) | 5.0 ns | 5.0 ns (×1.0) | 5.0 ns (×1.0) | 5.5 ns (×1.1) |
| Cycles/iteration | 53 k | 66 k (×**1.2**) | 54 k | 1440 k (×**26.6**) | 583 k (×**10.8**) | 165 k (×**3.0**) |
| Time/iteration | 264 us | 555 us (×**2.2**) | 270 us | 7200 us (×**26.6**) | 2915 us (×**10.8**) | 902 us (×**3.3**) |
| AT product | 5243 | 12,594 (×**2.4**) | 1880 | 37,892 (×**20.2**) | 15,043 (×**8.0**) | 5899 (×**3.1**) |

lower in both HLS designs. There is only a 20% overhead in terms of cycle count for both implementations of Lloyd's algorithm which indicates similar scheduling of operations. However, the HLS implementation has a significantly longer critical path (8.4 ns compared to 5.0 ns) which results in a performance gap of a factor of 2.1× in terms of latency and 2.4× in terms of AT product. The BRAM utilisation of the HLS design is lower because the synthesis tool decides to map some of the memories into LUT RAM. The last three columns show different variants of the HLS designs for the filtering algorithm. The design in Column 5 includes only code alterations that enable synthesisability and only uses Vivado's synthesis directives to improve QoR which results in a 20.2× degradation in terms of the AT product compared to the manual RTL design. Columns 6 and 7 show the importance of additional source-to-source transformations as discussed in Sect. 2.4.2. The manual partitioning of the heap memory narrows the performance gap from 20.2× to 8.0× (Column 6). The loop distribution in Listing 4 that enables pipelining in the tree traversal loop in addition to manual memory partitioning further improves the AT product to a factor of 3.1× larger than that of the manual RTL design (Column 7). The final AT product is more than two times smaller than that for Lloyd's algorithm.

## 2.6 Summary

This chapter presents a comparative case study for a C-to-FPGA flow using Xilinx Vivado HLS as an exemplary state-of-the-art tool. Our test cases are two alternative algorithms for $K$-means clustering, referred to as Lloyd's algorithm and the filtering algorithm. The former is a data flow-centric brute-force approach and has regular control flow and regular memory accesses, whereas the implementation of the filtering algorithm uses dynamic memory management and is based on recursive traversal of a pointer-linked tree structure. The search space pruning applied by the latter algorithm translates into a substantial run-time advantage in sequential software implementations. We first investigate the practicality of the algorithm in the context of an FPGA implementation and show that a carefully optimised parallel RTL implementation achieves the same execution time with four times fewer logic slices and three times fewer DSP slices. We also show how a custom implementation of dynamic memory allocation greatly reduces the on-chip memory consumption for the filtering algorithm. The implementations and evaluations of this part of the study were first published in [26].

The second part of this case study repeats the comparison for HLS designs of both algorithms. The performance gap between the HLS and hand-written RTL implementations of Lloyd's algorithm is approximately a factor of two in terms of area-time product, which is a remarkable result given the enormous difference in design time. The HLS design of the filtering algorithm also consumes a 'close-to-hand-written' amount of FPGA resources, but latency is initially degraded by a factor of 26.6×. The limited acceleration gained from semi-automatic design optimisations with synthesis directives results in a reversal of the previous finding: the AT product of the initial HLS implementation of the filtering algorithm is larger than that for Lloyd's algorithm. We subsequently apply manual code transformations to partition and privatise data structures accessed through pointers in order to promote parallelisation and to enable pipelining of the loop traversing the pointer-linked data structure which results in an overall 8× improvement of latency. The code transformations ultimately narrow the performance gap in terms of the AT product from 20.2× to 3.1× larger than that of the hand-crafted RTL design. The results of the HLS-based case study and guidelines for source code refactoring were first published in [27].

The AT product results in Table 2.3 show that both a carefully designed RTL and HLS implementation of the filtering algorithm outperform the respective implementation of the data flow-centric brute-force algorithm. This case study quantifies the benefits of hardware implementations of a sophisticated algorithm that uses structured data. We argue that this algorithm is representative of many other benchmarks that operate on tree structures, linked lists or graphs in general and common implementations of these algorithms are based on dynamically allocated data structures and pointer chasing. Due to the significant amount of source code refactoring in the implementation of the filtering algorithm, we conclude from this case study that the current generation of HLS tools lack support for effective design automation optimisations for this type of code. In particular, our code transformations enable memory partitioning, parallelisation and pipelining - optimisations that are essential for efficient FPGA designs. These optimisations require knowledge about data dependencies carried by data structures accessed through pointers.

Our goal in the following chapters of this thesis is to automate the memory partitioning and parallelisation in HLS flows targeting heap-manipulating programs. The difficult part of the automation of these optimisations is the program analysis: regardless of scope, every two heap-directed pointers could potentially reference the same memory cell and hence could create a data dependency. We propose an automated analysis of dependencies carried by data structures accessed through pointers, and an automated analysis to identify and privatise disjoint regions in the monolithic heap memory as the key features to improve the HLS support for (widely used) programs operating on dynamic, pointer-based data structures. Chapter 4 presents our approach to automatic heap partitioning and parallelisation. The HLS design aid in Chap. 4 automates the related code transformations that were applied manually in this chapter.

The synthesis of heap memory from on-chip BRAM in this case study and in Chap. 4 imposes a tight constraint on the working set size. For example, the RTL and HLS implementations in Sects. 2.3 and 2.4 use nearly 50% of the on-chip memory resources on the device. Chapter 5 removes this limitation by extending the technique in Chap. 4 to the automatic generation of application-specific parallel multi-cache systems in a framework where the heap resides in off-chip memory by default and only a fraction of it is held on-chip. This extension enables the HLS implementation of heap-manipulating programs with large memory footprints and alleviates the performance penalty due to the drop of memory bandwidth. Before describing the two core contributions of this thesis in Chaps. 4 and 5, we discuss related research in the following chapter.

## References

1. T. Kanungo, D. Mount, N. Netanyahu, C. Piatko, R. Silverman, A. Wu, An efficient K-means clustering algorithm: analysis and implementation. IEEE Trans. Pattern Anal. Mach. Intell. **24**(7), 881–892 (2002)
2. W. Meeus, K. Van Beeck, T. Goedemé, J. Meel, D. Stroobandt, An overview of todays high-level synthesis tools. Des. Autom. Emb. Syst., pp. 1–21 (2012)
3. Sobel Edge Detector. http://homepages.inf.ed.ac.uk/rbf/HIPR2/sobel.htm. Accessed 13 Mar 2016
4. S. Sarkar, S. Dabral, P. Tiwari, R. Mitra, Lessons and experiences with high-level synthesis. IEEE Des. Test Comput. **26**(4), 34–45 (2009)
5. BDTI, An Independent Evaluation of the AutoESL AutoPilot High-Level Synthesis Tool (2010). http://www.bdti.com/Resources/BenchmarkResults/HLSTCP/AutoPilot. Accessed 10 Oct 2012
6. R. Nane, V.-M. Sima, C. Pilato, J. Choi, B. Fort, A. Canis, Y.T. Chen, H. Hsiao, S. Brown, F. Ferrandi, J. Anderson, K. Bertels, A survey and evaluation of FPGA high-level synthesis tools, *IEEE Transactions on Computer-Aided Design of Integrated Circuits and Systems*. http://janders.eecg.toronto.edu/pdfs/tcad_hls.pdf. Accessed 28 Feb 2016
7. Y. Hara, H. Tomiyama, S. Honda, H. Takada, Proposal and quantitative analysis of the CHStone Benchmark program suite for practical C-based high-level synthesis. J. Inf. Process. **17**, 242–254 (2009)
8. A.K. Jain, Data clustering: 50 years beyond K-means. Pattern Recogn. Lett. **31**(8), 651–666 (2010)
9. C.M. Bishop, *Pattern Recognition and Machine Learning* (Springer, New York, 2006)
10. D. Clark, J. Bell, Multi-target state estimation and track continuity for the particle PHD filter. IEEE Trans. Aerosp. Electron. Syst. **43**(4), 1441–1453 (2007)
11. Y.-C. Hu, M.-G. Lee, K-means-based color palette design scheme with the use of stable flags. Electron. Imaging **16**(3), 033 003–033 003–11 (2007)
12. T. Saegusa, T. Maruyama, An FPGA implementation of real-time K-means clustering for color images. Real-Time Image Process. **2**(4), 309–318 (2007)
13. J.P. Theiler, G. Gisler, Contiguity-enhanced K-means clustering algorithm for unsupervised multispectral image segmentation. Proc. SPIE **3159**, 108–118 (1997)
14. M. Estlick, M. Leeser, J. Theiler, J.J. Szymanski, Algorithmic transformations in the implementation of K-means clustering on reconfigurable hardware, in *Proceedings of the ACM/SIGDA International Symposium on Field Programmable Gate Arrays (FPGA)* (2001), pp. 103–110
15. P. Drineas, A. Frieze, R. Kannan, S. Vempala, V. Vinay, Clustering large graphs via the singular value decomposition. Mach. Learn. **56**(1–3), 9–33 (2004)
16. M. Leeser, J. Theiler, M. Estlick, J. Szymanski, Design tradeoffs in a hardware implementation of the K-means clustering algorithm, in *Proceedings of the IEEE Sensor Array and Multichannel Signal Processing Workshop* (2000), pp. 520–524
17. X. Wang, M. Leeser, K-means clustering for multispectral images using floating-point divide, in *Proceedings of the IEEE International Symposium on Field-Programmable Custom Computing Machines (FCCM)* (2007), pp. 151–162
18. H.M. Hussain, K. Benkrid, A.T. Erdogan, H. Seker, Highly parameterized K-means clustering on FPGAs: comparative results with GPPs and GPUs, in *International Conference on Reconfigurable Computing and FPGAs* (2011), pp. 475–480
19. J. Kutty, F. Boussaid, A. Amira, A high speed configurable FPGA architecture for K-means clustering, in *Proceedings of the IEEE International Symposium on Circuits and Systems (ISCAS)* (2013), pp. 1801–1804
20. T.-W. Chen, S.-Y. Chien, Flexible hardware architecture of hierarchical K-means clustering for large cluster number. IEEE Trans. VLSI Syst. **19**(8), 1336–1345 (2011)
21. Vivado-KMeans: Hand-Written HDL Code and C-Based HLS Designs for K-means Clustering Implementations on FPGAs. https://github.com/FelixWinterstein/Vivado-KMeans. Accessed 19 Dec 2015

22. High-Level Synthesis with LegUp, Accessed 20 Oct 2015. http://legup.eecg.utoronto.ca/
23. ROCCC 2.0|Jacquard Computing, Accessed 12 May 2015. http://www.jacquardcomputing.com/roccc/
24. R. Nane, V. M. Sima, B. Olivier, R. Meeuws, Y. Yankova, K. Bertels, DWARV 2.0: A CoSy-based C-to-VHDL hardware compiler, in *Proceedings of the International Conference on Field Programmable Logic and Applications (FPL)* (2012), pp. 619–622
25. GAUT - High-Level Synthesis Tool From C to RTL, Accessed 21 Mar 2015. http://hls-labsticc.univ-ubs.fr/
26. F. Winterstein, S. Bayliss, G. Constantinides, FPGA-based K-means clustering using tree-based data structures, in *Proceedings International Conference on Field Programmable Logic and Applications (FPL)* (2013), pp. 1–6
27. F. Winterstein, S. Bayliss, G. Constantinides, High-level synthesis of dynamic data structures: a case study using Vivado HLS, in *Proceedings of the International Conference on Field-Programmable Technology (ICFPT)* (2013), pp. 362–365

# Chapter 3
# Background

Besides the basic HLS steps, resource allocation (the assignment of hardware components to operations), scheduling (the assignment of program operations to time slots), binding (assigning scheduled operations to functional units in the data path) and the generation of control circuits, an HLS tool usually performs several transformations of the input code. Many recent C-to-RTL flows build on standard compiler frameworks such as the *Low-Level Virtual Machine* (LLVM) compiler infrastructure [1] (e.g. Vivado HLS [2], ROCCC [3], LegUp [4], SDAccel [5] and the Altera SDK for OpenCL [6]) or GCC [7] (e.g. GAUT [8] and Bambu [9]). Especially recent HLS tools make use of the LLVM infrastructure, a popular framework which is used in many optimising software compilers. Within this framework, an input program is compiled into the *LLVM intermediate representation* (LLVM IR), an assembly-like language. Several high-level languages such as C/C++ or Java bytecode can be compiled into the IR using readily available front-ends such as `Clang` [10]. LLVM uses the *single static assignment* (SSA) form, i.e. every program variable is assigned exactly once. The SSA form results in explicit definition-usage (DEF-USE) chains in the IR, which simplifies some compiler optimisations. In an HLS tool, the IR passes through several standard compiler optimisations, for example dead-code elimination, constant propagation, loop unrolling, before hardware synthesis. The effect of standard LLVM optimisations on the QoR is explored in [11], where a 16% average improvement is reported.

This thesis offers a source-to-source compiler to improve the QoR of standard HLS tools that applies advanced HLS-specific code optimisations beyond standard software compiler optimisations. A crucial task during mapping a sequential program description into hardware is the extraction of parallelism while preserving the program semantics, which requires a dependence analysis. HLS flows usually apply standard compiler techniques to determine dependencies between program variables. However, detecting the absence data dependencies caused by aliasing of references to memory locations is a significantly more challenging task which is not supported for heap-directed pointers in standard HLS flows, as we demonstrated in the previous chapter. Additionally, parallelisation requires the memory system to match the computational parallelism. Compared to microprocessors, the distributed memory



F. Winterstein, *Separation Logic for High-level Synthesis*, Springer Thesis,
DOI 10.1007/978-3-319-53222-6_3

architecture in FPGAs provides an impressive memory bandwidth if the program data is partitioned and distributed over multiple on-chip memory blocks. Advanced C-to-FPGA compilers thus require a memory disambiguation for both parallelisation and memory partitioning. The objectives in this thesis are to implement a static program analysis and automated code transformations that enable automatic parallelisation, the distribution of data over separate blocks of on-chip memory, and the generation of parallel interfaces to external memory and parallel on-chip buffers.

The following literature review discusses two distinct approaches in an HLS context: unautomated approaches which rely on run-time profiling or manual source code annotations to determine data-level parallelism and approaches which use an automated framework for a static analysis, parallelisation and memory architecture generation. What follows is the discussion of limitations of these related approaches with respect to heap-manipulating programs as well as an introduction to separation logic, the theoretical framework leveraged in this thesis.

## 3.1 Profiling and User Annotation-Based Approaches

Cheng et al. [12] propose an HLS design aid targeting a hardware/software partitioned system consisting of a CPU and an FPGA accelerator which have both access to an external memory. Their technique generates an on-chip cache interface to an external memory using runtime profiling information. It consists of three phases. Firstly, the target application is profiled to identify independent partitions of memory accesses. Based on the profiling information, program operations accessing the same memory addresses are grouped into partitions and separate on-chip caches are assigned to disjoint on-chip memory regions accessed by the groups in a second step. Finally, a C-to-RTL flow generates the FPGA implementation of the accelerator with each program partition having access to its private cache. A fundamental advantage of the profiling-based approach is its versatility. However, runtime profiling requires a simulation environment on top of the HLS flow and a representative working set provided by the user to generate useful information. Furthermore, corner cases may be missed during simulation, i.e. identified partitions might still access memory addresses from other partitions and the generated hardware must be able to support these corner cases. Our approach in Chap. 5 is based on a static program analysis and therefore does not require simulation data.

Compiling C code to hardware targeting a CPU-FPGA architecture is also addressed in the CHiMPS framework [13]. The idea is similar in that it generates a parallel on-chip multi-cache (many-cache) architecture in order to feed parallel data paths. However, as opposed to the previous case, the identification of independent memory regions does not rely on profiling information but mainly on source code annotations with the `restrict`-keyword which states that two pointers do not access the same memory location. The fact that this aliasing information can be assumed to be exact ensures cache coherency among on-chip caches since a separate cache is created for any unique range of memory addresses. The exactness of the dependence information is thus beneficial in that it sidesteps coherency issues and additional overhead necessary to support corner cases. However, user intervention with manual source code annotations is required. As we shall see in the next chapter, many benchmarks with graph-traversing loops reach a state in which several loop iterations are independent of each other with respect to their memory accesses. This parallelisation opportunity firstly difficult to predict without a program analysis and secondly difficult to specify with code annotations. Another difference to our work is that shared memory regions are not supported by caches within the CHiMPS framework. We automatically insert a coherency network when it is required as we will describe in Chap. 5. Furthermore, the key difference to CHiMPS is our automated program analysis which allows our tool to parallelise the implementation make decisions as to when an expensive coherency mechanism is required and when it can be avoided.

While memory disambiguation is the main goal in this thesis, Sect. 5.4 of Chap. 5 describes application-specific cache sizing as an extension of our cache synthesis CAD flow. Recent related work has also explored the design space of the cache micro-architecture [14–16] beyond inter-cache coherency. Matthews et al. [15] explore the efficiency in terms of speed-up versus area increase of parallel coherent L1 caches with respect to size, associativity and replacement rule in an FPGA-based soft multi-core processor. Similarly, Choi et al. [16] compare different configurations of cache size, line size and associativity of shared on-chip caches, in addition to two approaches for increasing the number of access ports of the shared cache. The goal in this work is different: we infer cost/performance estimates prior to implementation and devise an automated cache system construction for a given application instead of exploring the cache micro-architecture. Automatic cache sizing from high-level specifications has been addressed in [13, 17]. Wingbermuehle et al. [17] implement a method similar to ours in that left-over memory resources are used to enhance the memory sub-system of stream-based kernels. Their work explores more parameters than our current technique (size, associativity, replacement rule and write policy), but the search in the parameter space is based on a simulated annealing-like technique. Another major difference to our work is that we target HLS applications without any assumption on the compute paradigm. CHiMPS' many-cache system [13] is notable in that it also constructs parallel caches based on left-over BRAM, clock rate degradation and predicted miss rate, although the prediction is not described in detail in the paper. The key difference of our work is the non-uniform sizing, which is realised by solving an optimisation problem to find the best assignment of cache sizes subject to a resource constraint.

## 3.2 Automated Static Analyses for Static Control Parts

Significant advancements in the direction of automated static analyses have been made for a specific type of loop kernels. Computation kernels in signal and image processing or scientific computing applications are often captured by `for`-loops or nests of `for`-loops. Their parallelisation therefore is a natural source for throughput improvement, which requires the memory system to support enough parallel data accesses. Many automated optimisations of the memory system in HLS literature focus on such loop-level optimisations and borrow many techniques originally developed for software compilers. In particular, they focus on a subclass of general loop nests, referred to as *static control parts* (SCoPs), where loop bounds and conditionals inside the loop are affine functions of the surrounding loop indices and constants (and possibly parameters). Array accesses within the loop body are likewise made through affine functions of the loop iterator variables. Due to the precise, static data dependence analysis that is possible for SCoPs, various transformations, such as loop tiling, loop splitting or merging, loop interchange or loop skewing, can be efficiently employed to promote loop- and memory-level parallelism or memory access optimisations in general. An underlying theoretical framework, which describes such an analysis and transformations in a unified mathematical abstraction, is referred to as the *polyhedral model* [18]. Because optimisations based on the polyhedral model are among the most popular advanced compiler techniques that have made their way into HLS CAD flows to date, we give a brief introduction here.

The polyhedral model is an algebraic representation of the execution of a program statement $S$ which is enclosed by an $n$-dimensional `for`-loop nest and conditionals. Such executions (in successive loop iterations) are denoted as dynamic instances of $S$. The bounds for the iteration variables of all enclosing loops as well as enclosing conditionals are affine functions of surrounding iterators, constants, and parameters. The iteration vector $x$ is an $n$-dimensional vector containing all surrounding loop iterators and each dynamic instance is associated with such an iteration vector. The set of all valid iteration vectors of a statement during execution of the loop nest spans a polytope (a bounded polyhedron) in $\mathbb{Z}^n$. This polytope can be represented by the set of $m$ linear inequalities describing the affine loop bounds and conditionals, i.e.

$$\mathcal{D} = \{x \in \mathbb{Z}^n | Ax \leq b\}, \tag{3.1}$$

where $A$ is an $m \times n$ matrix, $b$ is an $m$-dimensional vector, $m$ is the number of inequalities given by the loop bounds and conditionals, and the vector inequality $Ax \leq b$ is represented by the component-wise inequalities. If a loop nest depends on parameters which are modified by the program but remain constant during execution of the SCoP, a parameter vector can be added to the inequality in (3.1).

We assume memory accesses made by statement $S$ to be performed by references to an array $H$. For a precise analysis of these memory accesses, the array subscripts are usually an affine function of the iteration vector $x$, i.e.

$$g(x) = Fx + f, \tag{3.2}$$

where $g(x)$ and $f$ are $d$-dimensional vectors, $F$ is a $d \times n$ matrix, and $d$ is the dimensionality of array $H$. The data access function $g$ represents memory accesses at the granularity of array cells. Two statements are considered to be in dependence if both access the same memory cell (aliasing) and at least one of them performs a write access.

In addition to the iteration domain $\mathcal{D}$ and the data access $g$, an ordering of executions of $S$ must be modelled as a third aspect. Such an ordering is represented by the *scheduling function*, which associates each dynamic instance of $S$ with a logical date:

$$\Theta(x) = Tx + t, \tag{3.3}$$

where $\Theta(x)$ and $t$ are $k$-dimensional vectors, $T$ is a $k \times n$ matrix, and $k$ is the dimensionality of the logical date (time stamp). The ordering of logical dates is given by the lexicographic ordering (denoted as $\prec$) of $t$, i.e. $\Theta(x_1) \prec \Theta(x_2)$ means that dynamic instance $x_1$ is scheduled before $x_2$. In general, the ordering is not limited to temporal ordering but can also have a spatial meaning (e.g. scheduling iterations on different processors or different functional units on a chip) which is why the scheduling function is also referred to as the *scattering function* [19]. The scattering function $\Theta$ can apply a new lexicographic ordering to the original polyhedron $\mathcal{D}$. A description of such a transformation framework is given in [19].

These transformations, which represent a mathematical abstraction of loop transformations, are performed to improve performance, for instance to exhibit parallelism or improve data locality. Due to the possibility of statically analysing memory accesses, data dependencies between statements or loop iterations can be accounted for in the scheduling and it can be ensured that the program semantics are preserved.

There is a large body of work on code optimisations leveraging the polyhedral model in the domain of software compilers. The polyhedral model also became popular in an HLS context within the past decade. Liu et al. [20] have pioneered the use of the polyhedral model for inserting on-chip reuse buffers into the interface of an FPGA accelerator to an external memory. These reuse buffers hold data which are accessed by the loop kernel multiple times in order to reduce the number of slow accesses to the external memory. The polyhedral model is used to determine data reuse opportunities and to calculate the reuse volume at compile time. In [21], loop transformations are explored automatically in order to find a sequence of transformations that maximises parallelism and data locality. SCoPs are also targeted in [22] for an FPGA system which accesses data from an external synchronous DRAM (SDRAM). The memory architecture is optimised in two respects: in addition to the insertion of data reuse buffers, the number of unfavourable address sequences that cause time-consuming SDRAM row swaps is reduced by reordering the original address sequence. Data reuse and transaction reordering are based on data access analysis using the polyhedral model. The on-chip buffers in both cases differ from standard caches that are designed to work with arbitrary address sequences which are unknown at compile time. In SCoPs, exact compile-time knowledge about the data volume that is loaded into and fetched from a buffer in each iteration is available.

Cong et al. [23] implement bandwidth optimisations through memory partitioning based on a dependence analysis using an integer linear programming (ILP) formulation over the polyhedral model. Bondhugula et. al. [24] describe a scalable ILP-based technique for the aggregation of sets of loop iterations into tiles so as to maximise loop-level parallelism and data locality. Their technique is implemented in a source-to-source translator targeting code optimisations for FPGA-directed HLS [25].

## 3.3 Limitations and Extensions of the Polyhedral Framework

The polyhedral model is a powerful framework for automatic optimisation due its representation of optimisation sequences in a unified algebraic framework. It is, however, restricted to statically analysable loop-based program kernels as described in the previous section. This is a strong limitation as, in general, many programs do not strictly fulfil these requirements. There are several approaches aiming to remove this limitation in the context of the polyhedral model, most of which originate from the software compiler community. Relaxing the constraints of SCoPs mainly involves modelling loops other than `for`-loops with statically determinable bounds (such as `while`-loops), modelling arbitrary conditionals (such as data-dependent conditionals), analysing arbitrary memory accesses (such as indirect array references or heap-directed pointer accesses), or modelling loop nests depending parameters whose values are determined at run-time.

An approach to fit the polyhedral model to kernels with `while`-loops, arbitrary (non-statically determinable) conditionals in the loop body, and indirect array references is described by Benabderrahmane et al. [26]. A `while`-loop is transformed into a `for`-loop iterating from 0 to infinity, that is the iteration domain $\mathcal{D}$ of a loop becomes $\mathbb{N}$. The loop bound check is implemented with an exit predication which encloses the loop body statement and terminates the infinite loop with a `break`-statement. Arbitrary conditionals are implemented using *control predications* which individually predicate each statement enclosed by the conditional. Both exit and control predications are added to the iteration domain of a statement as additional constraints and predication evaluations are added to the loop body as additional statements. Benabderrahmane et al.'s analysis is a conservative over-approximation in that it assumes control predications to be true in all cases. Furthermore, if a statement contains an indirect array reference (such as a subscript of subscript) their dependence analysis conservatively assumes a dependency between this statement and every other statement accessing the same array. That is an array with such an access is considered a single scalar variable. The same holds for heap-manipulating statements. The extended framework allows them to perform a subset of standard code optimisations based on the polyhedral model on more irregular kernels than SCoPs. However, the approach is not suitable for the disambiguation of heap-directed pointer accesses.

Handling arbitrary loop bounds and conditionals, irregular memory accesses and run-time parameters in software compiler optimisations is also addressed by Jimborean et al. [27]. Instead of conservatively adapting the polyhedral model to fit to general loop kernels and performing static analysis, the assumption is that some dependence information, for instance in the case of indirect array references or pointer accesses, is not available at compile time. Thus the analysis phase is divided into a static and a dynamic part, while the latter fills in missing information after an online profiling phase. The run-time profiling monitors the first iterations of a loop nest and determines dependencies in a dynamic dependence analysis. The information is used to speculatively parallelise the loop nest by performing transformations based on the polyhedral model at runtime. The optimistic optimisation framework is based on a *thread level speculation system* which executes speculatively transformed code and provides a 'roll-back' mechanism in case of a wrong prediction. The technique can optimise loop kernels containing indirect array accesses, pointer-linked data structures (linked lists), arbitrary conditionals and `while`-loops. The heavy-weight online optimisation shares the same drawbacks with profiling-based memory architecture optimisation discussed in Sect. 3.1 in that the speculative optimisation has to account for mispredictions.

## 3.4 HLS Support for Pointers and Dynamic Memory Allocation

As discussed in the previous section, the polyhedral framework can be extended to support irregular control structures. The static analysis resorts to the overly pessimistic assumption that an indirect memory reference aliases with every other statement accessing the same array or every two pointers referencing a location in the heap alias. In addition, pointer-manipulating programs often use dynamic allocation and de-allocation of memory space during run-time. Dynamic memory allocation allows an application to request *just enough* memory required for its execution, leaving the remaining heap free. We have shown in Chap. 2 that the *just-enough* portion is up to 57× smaller than the worst-case. Dynamic memory allocation and heap-directed pointers are 'standard' features in software, including large bases of legacy codes. In particular, many data analytics algorithms in data center workloads are based on pointer-based data structures [28]. We argue that extending current HLS flows in order to support these is a large step towards HLS for full-featured C code.

The generation of currently available HLS tools [2–6, 8, 29–35], including Vivado HLS, avoid the issue of synthesising heap-directed pointers into hardware. There are several related research activities that seek to extend the support for heap-manipulating programs with dynamic memory allocation in contemporary FPGA-targeted HLS flows. Simsa et al. [36] describe a technique in which all heap operations (`new`, `delete` and pointer dereferencing) are translated into operations on a pre-allocated shared array and a global controller is included which keeps track of the free entries in the array. The size of this array is determined by static analysis by Cook et al. [37] that attempts to compute a parametric expression describing the maximum heap memory consumption. The parameters in this expression are program variables, so once their values are known, an absolute heap bound can be determined.

Our approach in Chap. 4 implements heap memory in the same way (instantiating arrays and turning pointer dereferencing into array accesses). However, the main difference is that our analysis breaks the monolithic heap into many disjoint portions that can be accessed in parallel. Furthermore, our extension in Chap. 5 places the heap in external memory (board-level DRAM and host-level main memory), supported by on-chip caches. This avoids the need for a compile-time analysis to determine heap bounds, which is not always possible. In addition, pre-allocating the maximum amount of memory can be a very conservative over-approximation of the amount required in the average case as we show in the previous chapter. Bambu [9], an academic HLS tool, uses a similar approach in that the program data is pre-characterized and the tool automatically decides for each data item whether it is stored in on-chip memory or whether an external memory interface is generated for it. Although not explicitly pointed out, we believe that this framework can also support dynamically allocated data.

The implementation of dynamic memory allocation is part of the necessary infrastructure of our work. However, the main contribution of this thesis is the automatic parallelisation of pointer-based programs and the partitioning of heap-allocated data structures across physically disjoint on-chip memories and buffers. Alias analyses have a long-standing tradition research on optimising software compilers [38–48], the majority of which collect a set of alias pairs, i.e. a pair of two pointer variables referring to the same location. Flow-sensitive analyses (taking the order of instruction executions into account) [38–40, 43, 48] provide better accuracy than flow-insensitive analyses [42, 44, 45, 47]. Accuracy can be measured in terms of the ability to rule out aliasing pairs that do not exist in reality but that an analysis must conservatively assume exist. Context-sensitivity in interprocedural analyses (taking the individual call-site information of a sub-routine into account) [40, 43, 48] further improves the analysis accuracy. The LLVM infrastructure also includes several such 'standard' alias analyses (including [45]). Heap-allocated recursive data structures are especially challenging for the analyses above because of the potentially unbounded number of aliases through the link fields and many of the techniques above cannot handle them. In [43], recursive data structure are treated as single cells, which excludes the possibility of partitioning. A technique called *k-limiting* [38, 47] determines the aliasing properties of the first $k$ elements of a linked structure ($k$ access paths), where $k$ is an arbitrary constant. However, this approximation provides no knowledge beyond the depth $k$.

The techniques in [41, 46] implement 'precise' analyses of the aliasing properties of recursive structures. Deutsch [41] uses a symbolic representation of access paths to reason about all elements in a data structure. The work by Ghiya and Hendren [46], in line with this work, uses a shape analysis of the heap layout to establish disjointness of heap-allocated recursive data structures for parallelising software compilers. This information is used to parallelise loops traversing these data structures, which is similar to one of our objectives. Their analysis classifies data structures into trees, lists, and general graphs and looks up the known aliasing properties of the link fields. Separation logic provides a more canonical approach to encoding the aliasing properties of data structures. For example, a separation logic-based analysis is aware that the memory portions allocated by two calls to `new` are disjoint and propagates this information through the execution trace of the program because the semantics of `new` and other heap-manipulating commands is embedded in the analysis. In principle, separation logic avoids the need for classifying data structures according to their aliasing properties. As we shall see in Sect. 3.5, our analysis also uses predicates for trees and linked list segments and other data structures in order to be able to analyse loops with unknown iterations count. If such a data structure is built up from data in disjoint memory portions, the aliasing properties can be automatically inferred as demonstrated by Guo et al. [49]. However, the key difference of our work to [46] is that we implement a heap memory footprint analysis which, besides proving the absence of data dependencies for program parallelisation, guides a hardware compiler to synthesise a distributed on-chip memory system, where data structure partitions reside in physically disjoint memory spaces. As we will discuss in Chap. 5, such a fine-grain footprint analysis is crucial when an implementations accesses a mixture of disjoint and shared memory resources.

Séméria et al. [50] present an approach for mapping C code with pointers and `malloc/free` operations into hardware and implement a distributed memory system. Similar to our work they instantiate on-chip allocator blocks using standard allocation schemes and use a pointer analysis to safely map the monolithic heap space to distributed on-chip memory banks. Their approach is based on a pointer analysis by Wilson and Lam [43] that uses a summary of different aliasing cases of the pointer arguments passed to a procedure to identify pointer-induced data dependencies. A fundamental difference to our approach is their approximate representations of data structures (*location sets* [43]), which can disambiguate accesses to different data structures, but does cannot partition recursive data structures. Our analysis precisely describes the shape of the heap layout. The approach to synthesis of pointer-based C code programs by Babb et al. [51] also uses an analysis based on location sets. In contrast to both, our approach allows us to partition recursive data structures, such as linked lists and trees, to increase parallelism.

The key differences of our work in the following Chap. 4 to the related above is the automated heap footprint analysis combined with the synthesis of a distributed memory architecture and automatic parallelisation of heap-manipulating code for hardware implementations. Our departure point from previous work on heap partitioning above is the use of recent advances in separation logic [52] which allows a formal description of the program state and reasoning about the resources accessed by a program. Because separation logic forms the central theoretical framework for this thesis, we give an introduction in the following section.

## 3.5 Static Analysis Based on Separation Logic

The introduction in this section discusses the fundamentals of separation logic and primarily targets readers who are non-experts in theoretical computer science. A more formal introduction to separation logic is given in [52].

The objective of our analysis is to identify disjoint regions in the heap memory that are accessed by different fragments of the program code so as to declare these code fragments as independent (given that no other dependencies exist). In our static analysis, we describe the layout of the heap with a formula at each point of program execution: Informally, it steps through the source code and maintains a formula describing the heap-allocated data structures as well as all points-to information at each program statement. While stepping (symbolically) from one statement to the next, the formula is modified reflecting the heap manipulation, for example a statement may allocate new data, dispose data, or change the data content. The formula maintains information about the layout of the data structure and ignores other properties such as their size. Thus, we refer to this type of analysis as *shape analysis*. Separation logic allows us to express the heap layout in concise formulae and to identify precisely what program statement accessed what part of the formula. The following sub-sections describe the required components of this analysis: the syntax of separation logic formulae (Sect. 3.5.1), the formal specification of program statements (Sect. 3.5.2), symbolically stepping through the source code (Sect. 3.5.3), and theorem proving in separation logic (Sect. 3.5.4), which informs us about the 'accessed' portion of the formula.

### 3.5.1 Modelling Program State in Separation Logic

A program modifies the values of program variables and the content of memory cells during execution. The assignment of values to variables and memory cells is referred to as *program state*. Separation logic is an extension of the *Hoare logic* [53]. It formally describes the program state with two components. The *store* describes the values assigned to variables (e.g. $x = 3$ means that variable $x$ currently holds the value 3) and the *heap* describes the values assigned to addressable memory locations (e.g. $y \mapsto 4$ means that pointer variable $y$ points to a memory cell containing the value 4). Note that $y \mapsto 4$ implies that the memory location at $y$ is allocated. A program may start with an empty heap memory where nothing is allocated, which is denoted by the *emp* keyword in separation logic formulae. In addition to program variables, the formulae may use auxiliary *primed* variables which only exist in formulae, not in the program code. For example, $z_1' = 4 \wedge y \mapsto z_1'$ means that there is some heap cell, containing the value 4 and $y$ points to that cell here, where '$\wedge$' is the classical 'and'-conjunction. The scope of $z_1'$ is bounded to the formula. The equation above is an abbreviation of $\exists z_1'.\ z_1' = 4 \wedge y \mapsto z_1'$. A primed variable is thus a placeholder for *some* value. For ease of readability, we omit the existential quantification ($\exists$) for primed variables in the remainder of this thesis.

Pointer variables can have a special value `nil` that corresponds to the `NULL` expression in C/C++. In addition to describing that a memory cell holds a scalar value, we can also use records (*structs* in C/C++): $y \mapsto [\mathtt{f}_1 : x_1', \ldots, \mathtt{f}_n : x_n']$ means that $y$ points to a heap-allocated record containing fields with $x_1', \ldots, x_n'$ as content. $\mathtt{f}_1, \ldots, \mathtt{f}_n$ are the field names.

Separation logic formulae are generally of the form $\Pi \wedge \Sigma$, where $\Pi$ is the *pure* part describing the store (e.g. $x = 3$) and $\Sigma$ is the *spatial* part describing the heap (e.g. $y \mapsto 4$). We define `Val` the set of values, `Var` the set of program variables, and `Var'` the set of auxiliary primed variables. Definition 3.1 defines the baseline syntax of the formulae used in our analysis.

**Definition 3.1** (*Baseline syntax of separation logic formulae*)

$$
\begin{array}{ll}
E, F ::= v \in \texttt{Val} \mid x \in \texttt{Var} \mid x'_i \in \texttt{Var}' & \text{expressions} \\
\Pi ::= \mathit{true} \mid E = F \mid E \neq F \mid \Pi \wedge \Pi & \text{pure formulae} \\
\Sigma ::= E \mapsto [\texttt{f}_1 : x'_1, \ldots, \texttt{f}_\texttt{n} : x'_n] \mid \mathit{emp} \mid \Sigma * \Sigma & \text{spatial formulae}
\end{array}
$$

Pure formulae contain (dis-) equalities and the classical conjunction ($\wedge$). Spatial formulae express the following:

- $E \mapsto [\texttt{f}_1 : x'_1, \ldots, \texttt{f}_\texttt{n} : x'_n]$ describes a heap-allocated record as discussed above. We use the abbreviation $E \mapsto \_$ to denote that $E$ points to 'some' record.
- *emp* denotes an empty heap where nothing is allocated.
- The separating conjunction ($*$) is the core element of separation logic: The formula $\Sigma_0 * \Sigma_1$ means that the heap is split into two disjoint portions $h_0$ and $h_1$, where $\Sigma_0$ holds for $h_0$ and $\Sigma_1$ holds for $h_1$. Disjoint heap portions are referred to as *heaplets*. The $*$-connective embeds the non-aliasing property of pointers, i.e. $E \mapsto [\texttt{f} : x'_1] * F \mapsto [\texttt{f} : y'_1]$ implies $E \neq F$ by definition. Hence, the content of the first heaplet can be modified by a program without any side effects for the second one. The usefulness of the separating conjunction becomes obvious when considering the counterexample in classical logic, $E \mapsto [\texttt{f} : x'_1] \wedge F \mapsto [\texttt{f} : y'_1]$: $E$ and $F$ may or may not alias, and expressing the non-aliasing property requires adding the constraint $E \neq F$ to the formula. These constraints are required for each pair of pointers in the program and quickly render an automated analysis unwieldy, especially in the case of pointer-linked data structures.

We refer to 'formula' as 'predicate' in the following. Definition 3.1 allows us to describe single, heap-allocated data records. To describe more sophisticated data structures such as linked lists or trees, we need to build additional predicates using the $*$-connective. For example, $E \mapsto [\texttt{n} : x'_1] * x'_1 \mapsto [\texttt{n} : x'_2]$ states that there exists a value $x'_1$ which occurs both in the `n`-field of the first record and is the address of the second record. Primed variables ($x'_1$ and $x'_2$) are useful here because they express the pointer link between two records without the need for knowing the physical address value of the link field ($x'_1$). A naive approach of describing a linked list is to mention all nodes in the list: $E \mapsto [\texttt{n} : x'_1] * x'_1 \mapsto [\texttt{n} : x'_2] * \ldots * x'_m \mapsto [\texttt{n} : \texttt{nil}]$. This, however, is problematic as the length $m$ of a dynamically allocated linked list is usually unknown at compile time. Instead, we use recursive predicates that describe data structures without knowing their size:

**Definition 3.2** (*Example: List segment*)

$$ls(E,F) \Longleftrightarrow (E = F \wedge emp) \vee (E \neq F \wedge E \mapsto [\mathtt{n}: x_1'] * ls(x_1', F)) \quad (3.4)$$

i.e. there is a list segment between pointer $E$ and $F$ if and only if the following condition holds. If $E = F$ this heap portion is empty. Otherwise $E$ points to an element which, in turn, points to a list segment between itself and $F$.

**Definition 3.3** (*Example: Tree*)

$$tree(E) \Longleftrightarrow (E = \mathtt{nil} \wedge emp) \vee (E \mapsto [\mathtt{l}: x_1', \mathtt{r}: y_1'] * tree(x_1') * tree(y_1')) \quad (3.5)$$

i.e. there is a tree pointed to by $E$ if and only if the following condition holds. If $E \neq \mathtt{nil}$ it points to an element which contains pointers to left and right sub-tree.

**Definition 3.4** (*Example: List with pointers to other heaplets*)

$$pls(E,F) \Longleftrightarrow (E = F \wedge emp) \vee$$
$$(E \neq F \wedge E \mapsto [\mathtt{u}: u_1', \mathtt{c}: c_1', \mathtt{n}: n_1'] * tree(u_1') * c_1' \mapsto \_ * pls(n_1', F)) \quad (3.6)$$

i.e. there is a list segment as in (3.4) whose elements also point to a tree and a heap-allocated record.

Note that we omitted additional data fields in the records above for ease of illustration. The above examples demonstrate the ability to describe common data structures; automatic inference of such definitions has been demonstrated by Guo et al. in [49].

### 3.5.2 Programming Language

The next step is to define how program state, expressed in separation logic formulae, is modified during program execution. For didactic purposes, we consider a simple programming language with heap-manipulating commands and loops:

**Definition 3.5** (*Programming language*)

| | |
|---|---|
| $b ::= E = F \mid E \neq F$ | boolean expressions |
| $A ::= x := E \mid x := [E].\mathtt{f} \mid [E].\mathtt{f} := F \mid$ new(x) $\mid$ delete(E) | atomic commands |
| $C ::= A \mid$ **if** $b\ C_1\ C_2 \mid$ **while** $b\ C \mid C_1; C_2$ | commands |

$E$ and $F$ are arbitrary expressions containing program variables and values (e.g. $E ::= x$, $E ::= \mathtt{nil}$, or $E ::= y + 1$). The term $[E].\mathtt{f}$ denotes pointer dereferencing of $E$ and accessing field $\mathtt{f}$ of the heap-allocated record pointed to by $E$.

The program statements (commands) modify the state. The transition of state upon execution of a command is specified by the triple $\{P\}C\{Q\}$. $P$ is the formula describing the pre-condition the state must satisfy for the command to run. If $C$ runs and halts then the post-condition formula $Q$ for the program state is true after execution [52]. For example, if $C$ is a command that writes the value 5 to the memory cell referenced by $y$ this heap cell must be allocated (pre-condition) and must contain 5 after successful command execution (post-condition): $\{y \mapsto [\mathtt{f} : x_1']\}\ [y].\mathtt{f} := 5\ \{y \mapsto [\mathtt{f} : 5]\}$. Definition 3.6 specifies a triple for each atomic command of our programming language:

**Definition 3.6** (*Specifications for atomic commands* [54])

$$\begin{array}{c}
\{\, x = y_1' \,\}\ x := E \qquad \{\, x = E[y_1'/x] \,\} \\
\{\, E \mapsto [\mathtt{f} : y_1'] \,\}\ [E].\mathtt{f} := F\ \{\, E \mapsto [\mathtt{f} : F] \,\} \\
\{\, x = y_1' \wedge E \mapsto [\mathtt{f} : z_1'] \,\}\ x := [E].\mathtt{f}\ \{\, x = z_1' \wedge E[y_1'/x] \mapsto [\mathtt{f} : z_1'] \,\} \\
\{\, emp \,\}\ new(x) \qquad \{\, x \mapsto z_1' \,\} \\
\{\, E \mapsto y' \,\}\ delete(E)\ \ \{\, emp \,\}
\end{array}$$

The term $E[y_1'/x]$ denotes expression $E$ with all occurrences of $x$ replaced by $y_1'$. Note that specifying pointer-manipulating commands in this way is only possible thanks to separation logic's *frame rule*. We discuss the frame rule in Sect. 3.5.3 below.

### 3.5.3 Symbolic Execution of Programs

Our static analysis 'symbolically' executes the program by propagating the program state, expressed in separation logic formulae, from one program statement to the next, thereby updating it using the specifications for single commands in Definition 3.6. We build our automated analysis on `coreStar` [55], which, in its original form, is a separation logic-based software verification tool. The tool includes a symbolic execution engine and a theorem prover. We discuss both components in this and the following section.

The symbolic execution propagates the state formula through all control flow paths of the program (branching and loops create multiple control flow paths). At each node in the control flow graph (CFG), `coreStar` determines the part of the formula describing the current state which matches the pre-condition of the current program statement, and replaces that part with the post-condition in Definition 3.6. The other parts, $F$, of the state formula remain untouched. Formally, before executing the program statement $C$, it breaks the current program state $\Pi_1 \wedge \Sigma$ into $\Pi_1 \wedge P * F$, where $P$ is the pre-condition of $C$ and $F$ is called the *frame*. The symbolic execution of $C$ then updates the program state to $\Pi_2 \wedge Q * F$ by replacing $P$ by $Q$ and leaving the frame $F$ untouched. The central idea of a separation logic-based symbolic execution is thus to consider a heap portion separately from its frame. Separation logic's *frame rule* formalises this behaviour. The frame rule is an *inference rule* of the form

$$\frac{\text{premise}}{\text{conclusion}}.$$

An inference rule asserts that "if the premise holds then the conclusion holds". The frame rule defines the invariance of the unmodified frame $F$ using the separating conjunction:

$$\frac{\{P\}C\{Q\}}{\{P * F\}C\{Q * F\}}, \text{ if } C \text{ does not modify any free variables in } F. \tag{3.7}$$

Separation logic thus provides a mechanism for a fine-grain analysis of the heap layout and to reason locally about the portion manipulated by a command while declaring the remaining memory cells unchanged. As opposed to classical program proving, local reasoning makes the analysis of pointer-manipulating programs tractable. Note that, in a 'correct' program, the symbolic execution always finds a suitable $P$, whereas failure to do so allows a software verification tool (e.g. [56]) to find a potential pointer-related bug. Here, we use separation logic for proving parallelisability instead of correctness, but, as a side effect, our tool also reports a failure in this case.

Our analysis in Chap. 4 uses our version of `coreStar` that we have modified to include an extension of the standard symbolic execution called *labelled symbolic execution* by Raza et al. [54]. This technique assigns a unique label to $Q$, the spatial part of the state formula that was modified, i.e. $\Pi_2 \wedge \Sigma \equiv \Pi_2 \wedge \langle Q \rangle_{\{l \in \texttt{Lab}\}} * F$, with `Lab` being the set of all labels. In the original work in [54], each program statement $C_i$ is assigned a unique label $l_i \in \texttt{Lab}$. The technique solves the problem that a state formula only describes the instantaneous state at a given point of execution, but does not describe the relations with formulae at other points of execution. The labels attached to the heaplets provide this link. For example, consider three program statements [x].f := 4, [y].f := 5 and [z].f := [x].f in the toy language from Definition 3.5 which are executed in sequence. The state formula at each point of labelled symbolic execution is shown in blue and, at the start, all label sets are empty:

$$\begin{aligned}
&\langle x \mapsto [\texttt{f} : x_1']\rangle_{\{\}} * \langle y \mapsto [\texttt{f} : y_1']\rangle_{\{\}} * \langle z \mapsto [\texttt{f} : z_1']\rangle_{\{\}} \\
1 : \ &[\texttt{x}].\texttt{f} := 4; \\
&\langle x \mapsto [\texttt{f} : 4]\rangle_{\{1\}} * \langle y \mapsto [\texttt{f} : y_1']\rangle_{\{\}} * \langle z \mapsto [\texttt{f} : z_1']\rangle_{\{\}} \\
2 : \ &[\texttt{y}].\texttt{f} := 5; \\
&\langle x \mapsto [\texttt{f} : 4]\rangle_{\{1\}} * \langle y \mapsto [\texttt{f} : 5]\rangle_{\{2\}} * \langle z \mapsto [\texttt{f} : z_1']\rangle_{\{\}} \\
3 : \ &[\texttt{z}].\texttt{f} := [\texttt{x}].\texttt{f}; \\
&\langle x \mapsto [\texttt{f} : 4]\rangle_{\{1,3\}} * \langle y \mapsto [\texttt{f} : 5]\rangle_{\{2\}} * \langle z \mapsto [\texttt{f} : 4]\rangle_{\{3\}}
\end{aligned}$$

The unique labels $l_i$ are the code lines of the statements here. The label sets are filled once a statement accessed the respective heaplet. The final label sets tell us that statements 1 and 3 accessed the same memory location, while the access of statement 2 is independent of the others. Raza's technique thus propagates the 'heap footprint' of each statement through the CFG. This tracks the memory accesses made by different parts of the program, a prerequisite for detecting heap-carried dependencies.

As we will describe in Chap. 4, our heap access analysis is a modified version of Raza's labelled symbolic execution. The main difference is that we embed different information in the label sets in order to detect the presence of communication-free parallelism in loops and to generate an assignment of heap partitions to physically distributed memories.

### 3.5.4 Theorem Proving

Automated theorem proving is the work horse in our tool flow. The symbolic execution engine uses it to infer the frame portion $F$ at each CFG node as described above. A detailed description of frame inference is beyond the scope of this introduction, but is given in [57]. It is also used to prove implications described in the next chapter. In all cases, the theorem prover tries to verify an *entailment* of the form $S_1 \vdash S_2$ which is interpreted as "$S_1$ entails $S_2$" or "from $S_1$ I can derive $S_2$", with $S_1$ and $S_2$ being formulae in separation logic of the form $\Pi \wedge \Sigma$. The theorem prover in `coreStar` builds on the proof technique in [57]. The basic idea is to reduce an entailment $S_1 \vdash S_2$ to an axiom $\Pi \wedge emp \vdash \texttt{true} \wedge emp$, with an arbitrary pure formula $\Pi$. The proof of the original entailment is successful if the reduction is successful. The entailment reduction is performed by applying a sequence of inference rules. Besides the frame rule, a separation logic theorem prover 'knows' a set of other inference rules. The proof engine in `coreStar` is generic in that it, except for the basic terms `true`, `false` and *emp*, (dis-) equalities, the separating conjunction and some general basic rules, no predicates are pre-defined; the user defines the underlying proof logic in a set of inference rules. The prover processes two types of rules: proof rules and abstraction rules. The former are used during the proof search for confirming the validity of an entailment: the theorem prover applies its proof rules upwards, i.e. the premise of the previous rule application becomes the conclusion of the current rule application until an axiom is reached or a contradiction is found. A proof rule modifies an entailment. For example, we can inform the prover that the following entailment is valid:

$$x \mapsto [\texttt{n} : x_1'] * ls(x_1', \texttt{nil}) \vdash ls(x, \texttt{nil}) \tag{3.8}$$

(i.e. if$x$ *points to the first element in a linked list, then*$x$ *itself points to a linked list)*. To this end, the prover needs two proof rules:

$$\frac{ls(E, F) \vdash ls(E, F)}{E \mapsto [\texttt{n} : x_1'] * ls(x_1', F) \vdash ls(E, F)} \qquad \frac{Q_1 \vdash Q_2}{Q_1 * S \vdash Q_2 * S} \tag{3.9}$$

The first rule simplifies the entailment by 'rolling up' the first element and the tail of a list segment into a list segment. The second is a 'subtraction rule' that removes identical heaplets on both sides of the entailment. Given these rules, the theorem prover will derive

$$\cfrac{\cfrac{emp \vdash emp}{ls(x, \texttt{nil}) \vdash ls(x, \texttt{nil})}\text{subtraction}}{x \mapsto [\texttt{n} : x_1'] * ls(x_1', \texttt{nil}) \vdash ls(x, \texttt{nil})}\text{roll} - \text{up} \tag{3.10}$$

Starting from the initial state $E \mapsto [\texttt{n} : x_1'] * ls(x_1', F) \vdash ls(E, F)$ in the bottom row, (3.10) shows the application of both inference rules in (3.9) from bottom to top. The top row is equivalent to $\texttt{true} \wedge emp \vdash \texttt{true} \wedge emp$ which is an axiom. Hence, (3.10) tells us that (3.8) can be derived from an axiom and therefore is a valid entailment.

The second class of rules are abstraction rules. The purpose of abstraction rules is to syntactically rewrite the current state formula $\Pi \wedge \Sigma$ that is propagated from statement to statement during the symbolic execution. Abstraction absorbs singleton heaplets in recursive predicates such as those in Definitions 3.2–3.4. For example, a formula can be rewritten so that the head node of a linked list and the tail list can be merged into one linked list. Formally, in $s \mapsto [\texttt{n} : x_1'] * ls(x_1', \texttt{nil})$, we can fold $s \mapsto [\texttt{n} : x_1']$ into the $ls$ predicate resulting in $ls(s, \texttt{nil})$. This rewrite step is called abstraction because we lose some information here: Instead of knowing that the heap contains a linked list with at least one entry, we now know that it contains a linked list which possibly can be empty. However, the information of having at least one node in the list is not required by our analysis because we are interested in the shape of the heap layout only. As we shall see in Chap. 4, abstraction plays a critical role in our loop analysis. We maintain a set of abstraction rules which we provide to the theorem prover and which define what is a valid abstraction. Abstraction rules are of the form:

$$\frac{\text{condition}}{\Sigma' \rightsquigarrow \Sigma''} \tag{3.11}$$

The rule is applied as soon as the condition holds and rewrites the spatial part of a formula $\Pi \wedge \Sigma$ with $\Sigma \equiv \Sigma' * \Sigma_F$. $\Sigma'$ is replaced by $\Sigma''$. $\Sigma_F$ is an arbitrary *context* in $\Sigma$ which is preserved by the rewrite rule. Our abstraction rules define when the analysis is allowed to fold singleton heaplets into the recursive predicates of Definitions 3.2–3.4. We adopt the approach of Magill et al. [58] for defining the condition as to when folding occurs: Our abstraction rules allow folding across primed variables, but forbid folding across program variables, e.g. $s \mapsto [\texttt{n} : x_1'] * ls(x_1', \texttt{nil})$ is folded into $ls(s, \texttt{nil})$, but $s \mapsto [\texttt{n} : x] * ls(x, \texttt{nil})$ does not get merged into $ls(s, \texttt{nil})$ because $x$ is a program variable. The following abstraction rules for $ls$, $tree$ and $pls$ predicates formalise this condition. In addition to Magill's technique, our rules define how heap footprint labels are affected by the folding operation:

**Definition 3.7** (*Basic abstraction rules for ls predicates*)

$$\frac{x_1' \notin context \cup \{E\}}{\langle E \mapsto [\mathtt{n} : x_1']\rangle_{\mathtt{Lab1}} * \langle x_1' \mapsto [\mathtt{n} : F]\rangle_{\mathtt{Lab2}} \rightsquigarrow \langle ls(E, F)\rangle_{\mathtt{Lab1} \cup \mathtt{Lab2}}}$$

$$\frac{x_1' \notin context \cup \{E\}}{\langle E \mapsto [\mathtt{n} : x_1']\rangle_{\mathtt{Lab1}} * \langle ls(x_1', F)\rangle_{\mathtt{Lab2}} \rightsquigarrow \langle ls(E, F)\rangle_{\mathtt{Lab1} \cup \mathtt{Lab2}}}$$

$$\frac{x_1' \notin context \cup \{E\}}{\langle ls(E, x_1')\rangle_{\mathtt{Lab1}} * \langle ls(x_1', F)\rangle_{\mathtt{Lab2}} \rightsquigarrow \langle ls(E, F)\rangle_{\mathtt{Lab1} \cup \mathtt{Lab2}}}$$

i.e. two list nodes, a list node and a list tail, or two list segments are folded into a single list, respectively. The rule fires if the linking pointer is a primed variable and appears nowhere else in the current state formula. The set of footprint labels attached to a predicate resulting from merging two predicates is the union of both original label sets ($\mathtt{Lab1} \cup \mathtt{Lab2}$).

**Definition 3.8** (*Basic abstraction rule for tree predicates*)

$$\frac{x_1', y_1' \notin context \cup \{E\}}{\langle E \mapsto [\mathtt{l} : x_1', \mathtt{r} : y_1']\rangle_{\mathtt{Lab1}} * \langle tree(x_1')\rangle_{\mathtt{Lab2}} * \langle tree(y_1')\rangle_{\mathtt{Lab3}} \rightsquigarrow \langle tree(E)\rangle_{\mathtt{Lab1} \cup \mathtt{Lab2} \cup \mathtt{Lab3}}}$$

i.e. a tree node and both sub-trees are folded into a single tree. The rule fires if the linking pointers are primed variables and appears nowhere else in the current state formula. The set of footprint labels attached to a predicate resulting from merging two predicates is the union of both original label sets ($\mathtt{Lab1} \cup \mathtt{Lab2} \cup \mathtt{Lab3}$).

**Definition 3.9** (*Basic abstraction rules for pls predicates*)

$$\frac{u_1', c_1', n_1' \notin context \cup \{E\}}{\langle E \mapsto [\mathtt{u} : u_1', \mathtt{c} : c_1', \mathtt{n} : n_1']\rangle_{\mathtt{Lab1}} * \langle tree(u_1')\rangle_{\mathtt{Lab2}} * \langle c_1' \mapsto \_\rangle_{\mathtt{Lab3}} * \langle pls(n_1', F)\rangle_{\mathtt{Lab4}} \rightsquigarrow \langle pls(E, F)\rangle_{\mathtt{Lab1} \cup \mathtt{Lab2} \cup \mathtt{Lab3} \cup \mathtt{Lab4}}}$$

$$\frac{x_1' \notin context \cup \{E\}}{\langle pls(E, x_1')\rangle_{\mathtt{Lab1}} * \langle pls(x_1', F)\rangle_{\mathtt{Lab2}} \rightsquigarrow \langle pls(E, F)\rangle_{\mathtt{Lab1} \cup \mathtt{Lab2}}}$$

i.e. a list node and a list tail, or two list segments are folded into a single list, respectively. The rule fires if the linking pointers are primed variables and appear nowhere else in the current state formula. The set of footprint labels attached to a predicate resulting from merging two predicates is the union of all original label sets. The folding of two list nodes and their linked predicates is omitted here for ease of readability, but is analogous to the first rule of Definition 3.7.

Note that, for ease of explanation, Definition 3.7–3.9 only show a subset of the abstraction rules used by our analysis.

### 3.5.5 *Application to HLS*

Formal software verification has been the main application of separation logic. Only recently, its scope has been extended to data dependence analyses for automatic parallelisation. Raza et al. [54] use their labelled symbolic execution and heap footprint analysis for an analysis of pointer-induced dependencies enabling the parallelisation of software programs. We build on the labelled symbolic execution framework, but our analysis embeds different information in the heaplet label sets as we shall see in the next chapter. We also extend their method by allowing the analysis to perform semantics-preserving modifications to the program state until the partitioning goal can be proven. Another difference is that we propose an analysis tailored to loop parallelisation and the inference of loop-invariant state descriptions which is not covered in [54]. In contrast, our analysis searches eagerly for parallelisation opportunities in the iteration space of pointer-chasing loops and makes code transformations on-the-fly to achieve its goal.

The work in [59] is notable in that it also takes Raza's method into an HLS context. The parallelisation transformations, however, are not automated and memory partitioning is not addressed. Furthermore, determining disjointness in our tree-based benchmarks requires successive unrollings of loop iterations before disjointness can be established, which is not implemented in their technique. Finally, concurrent work by Botinčan et al. [60] describes a technique for separation logic-based parallelisation of software threads. Their work is interesting in that they automatically insert synchronisation to preserve dependencies in addition to a dependence analysis, a feature that we also add to the extension of our baseline technique (Chap. 5). Their work, however, focuses on the theoretical framework whereas we use the theoretical foundations in a demonstrably practical implementation. Furthermore, our tool seamlessly connects to existing HLS flows because the analysis and code transformation operates on LLVM IR. Villard develops `llstar` [61], a tool that uses similar specifications of LLVM bitcode instructions in separation logic. The major difference to our work is that `llstar` is a software verification tool that aims to find pointer-related bugs (e.g. `NULL`-dereferencing) in a program, whereas we link our heap analysis and bitcode specifications with source-to-source transformations for memory partitioning in hardware implementations. To the best of our knowledge, our work is the first separation logic-inspired tool that automatically synthesises a distributed memory system and parallel hardware implementations form heap-manipulating code in the context of state-of-the-art HLS flows.

## References

1. The LLVM Compiler Infrastructure. http://llvm.org/. Accessed 28 Feb 2016
2. Xilinx Vivado HLS, Accessed 12 May 2015. http://www.xilinx.com/products/design-tools/vivado/integration/esl-design.html

3. ROCCC 2.0|Jacquard Computing, Accessed 12 May 2015. http://www.jacquardcomputing.com/roccc/
4. High-Level Synthesis with LegUp, Accessed 20 Oct 2015. http://legup.eecg.utoronto.ca/
5. Xilinx SDAccel Development Environment for OpenCL, Accessed 13 Jan 2016. http://www.xilinx.com/products/design-tools/software-zone/sdaccel.html
6. Altera SDK for OpenCL, Accessed 13 Jan 2016. https://www.altera.com/products/design-software/embedded-software-developers/opencl/overview.html
7. GCC, the GNU Compiler Collection. https://gcc.gnu.org. Accessed 21 Aug 2014
8. GAUT - High-Level Synthesis Tool From C to RTL, Accessed 21 Mar 2015. http://hls-labsticc.univ-ubs.fr/
9. C. Pilato, F. Ferrandi, Bambu: a modular framework for the high level synthesis of memory-intensive applications, in *Proceedings International Conference on Field Programmable Logic and Applications (FPL)* (2013), pp. 1–4
10. Clang: A C Language Family Frontend for LLVM. http://clang.llvm.org. Accessed 28 Feb 2016
11. Q. Huang, R. Lian, A. Canis, J. Choi, R. Xi, S. Brown, J. Anderson, The effect of compiler optimizations on high-level synthesis for FPGAs, in *Proceedings of the IEEE International Symposium on Field-Programmable Custom Computing Machines (FCCM)* (2013), pp. 89–96
12. S. Cheng, M. Lin, H.J. Liu, S. Scott, J. Wawrzynek, Exploiting memory-level parallelism in reconfigurable accelerators, in *Proceedings of the IEEE International Symposium on Field-Programmable Custom Computing Machines (FCCM)* (IEEE, New Jersey, 2012), pp. 157–160
13. A. Putnam, S. Eggers, D. Bennett, E. Dellinger, J. Mason, H. Styles, P. Sundararajan, R. Wittig, Performance and power of cache-based reconfigurable computing. ACM SIGARCH Comput. Architect. News **37**(3), 395 (2009)
14. H.-J. Yang, K. Fleming, M. Adler, F. Winterstein, J. Emer, Scavenger: automating the construction of application-optimized memory hierarchies, in *Proceedings of the International Conference on Field Programmable Logic and Applications (FPL)* (2015), pp. 1–8
15. E. Matthews, N.C. Doyle, L. Shannon, Design space exploration of L1 data caches for FPGA-based multiprocessor systems, in *Proceedings of the ACM/SIGDA International Symposium on Field-Programmable Gate Arrays (FPGA)* (2015), pp. 156–159
16. J. Choi, K. Nam, A. Canis, J. Anderson, S. Brown, T. Czajkowski, Impact of cache architecture and interface on performance and area of FPGA-based processor/parallel-accelerator systems, in *Proceedings of the IEEE International Symposium on Field-Programmable Custom Computing Machines (FCCM)* (2012), pp. 17–24
17. J.G. Wingbermuehle, R.K. Cytron, R.D. Chamberlain, Superoptimized memory subsystems for streaming applications, in *Proceedings of the ACM/SIGDA International Symposium on Field-Programmable Gate Arrays (FPGA)* (2015), pp. 126–135
18. P. Feautrier, Dataflow analysis of array and scalar references. Int. J. Parallel Program. **20**(1), 23–53 (1991)
19. C. Bastoul, Code generation in the polyhedral model is easier than you think, in *Proceedings of the International Conference on Parallel Architectures and Compilation Techniques* (2004), pp. 7–16
20. Q. Liu, G.A. Constantinides, K. Masselos, P.Y. Cheung, Automatic on-chip memory minimization for data reuse, in *Proceedings of the IEEE International Symposium on Field-Programmable Custom Computing Machines* (IEEE, New Jersey, 2007), pp. 251–260
21. H. Devos, K. Beyls, M. Christiaens, J. Van Campenhout, E.H. D'Hollander, D. Stroobandt, Finding and applying loop transformations for generating optimized fpga implementations. Trans. High Perform. Embed. Architect. Compil. I **4050**, 159–178 (2007)
22. S. Bayliss, G. Constantinides, Optimizing SDRAM bandwidth for custom FPGA loop accelerators, in *Proceedings of the ACM/SIGDA International Symposium on Field Programmable Gate Arrays (FPGA)* (ACM Press, 2012), pp. 195–204
23. J. Cong, W. Jiang, B. Liu, Y. Zou, Automatic memory partitioning and scheduling for throughput and power optimization. ACM Trans. Des. Autom. Electron. Syst. **16**(2), 1–25 (2011)

24. U. Bondhugula, A. Hartono, J. Ramanujam, P. Sadayappan, A practical automatic polyhedral parallelizer and locality optimizer. SIGPLAN Notices **43**(6), 101–113 (2008)
25. L.-N. Pouchet, P. Zhang, P. Sadayappan, J. Cong, Polyhedral-based data reuse optimization for configurable computing, in *Proceedings of the ACM/SIGDA International Symposium on Field Programmable Gate Arrays (FPGA)* (ACM, 2013), pp. 29–38
26. M. Benabderrahmane, L. Pouchet, A. Cohen, C. Bastoul, The polyhedral model is more widely applicable than you think, in *Proceedings of the International Conference on Compiler Construction* (2010), pp. 283–303
27. A. Jimborean, Adapting the Polytope Model for Dynamic and Speculative Parallelization, Ph.D. Thesis, Université de Strasbourg (2012). http://tel.archives-ouvertes.fr/tel-00733850
28. G. Weisz, J. Melber, Y. Wang, K. Fleming, E. Nurvitadhi, J.C. Hoe, A study of pointer-chasing performance on shared-memory processor-FPGA systems, in *Proceedings of the ACM/SIGDA International Symposium on Field-Programmable Gate Arrays* (2016), pp. 264–273
29. Calypto Catapult Synthesis, Accessed 19 Dec 2015. http://calypto.com/en/products/catapult/overview/
30. Cadence C-to-Silicon Compiler, Accessed 23 Dec 2015. http://www.cadence.com/products/sd/silicon_compiler/
31. Cadence Cynthesizer Solution, Accessed 23 Dec 2015. http://www.cadence.com/products/sd/cynthesizer/
32. Cadence Stratus High-Level Synthesis, Accessed 22 Dec 2015. http://www.cadence.com/products/sd/stratus/
33. Impulse CoDeveloper, Accessed 25 Nov 2015. http://www.impulseaccelerated.com/products.htm
34. Synopsys Synphony C Compiler, Accessed 26 Nov 2015. https://www.synopsys.com/Tools/Implementation/RTLSynthesis/Pages/SynphonyC-Compiler.aspx
35. R. Nane, V.M. Sima, B. Olivier, R. Meeuws, Y. Yankova, K. Bertels, DWARV 2.0: a CoSy-based C-to-VHDL hardware compiler, in *Proceedings of the International Conference on Field Programmable Logic and Applications (FPL)* (2012), pp. 619–622
36. J. Simsa, S. Singh, Designing hardware with dynamic memory abstraction, in *Proceedings of the ACM/SIGDA International Symposium on Field Programmable Gate Arrays (FPGA)* (2010), pp. 69–72
37. B. Cook, A. Gupta, S. Magill, A. Rybalchenko, J. Simsa, S. Singh, V. Vafeiadis, *Finding Heap-Bounds for Hardware Synthesis, in Formal Methods in Computer-Aided Design* (IEEE, New York, 2009), pp. 205–212
38. W. Landi, B.G. Ryder, A safe approximate algorithm for interprocedural aliasing, in *Proceedings of the ACM SIGPLAN Conference on Programming Language Design and Implementation* (1992), pp. 235–248
39. J.-D. Choi, M. Burke, P. Carini, Efficient flow-sensitive interprocedural computation of pointer-induced aliases and side effects, in *Proceedings of the ACM SIGPLAN-SIGACT Symposium on Principles of Programming Languages* (1993), pp. 232–245
40. M. Emami, R. Ghiya, L.J. Hendren, Context-sensitive interprocedural points-to analysis in the presence of function pointers, in *Proceedings of the ACM SIGPLAN Conference on Programming Language Design and Implementation* (1994), pp. 242–256
41. A. Deutsch, Interprocedural May-alias analysis for pointers: beyond K-limiting. SIGPLAN Notices **29**(6), 230–241 (1994)
42. M. Burke, P. Carini, J.-D. Choi, M. Hind, *Proceedings of the International Workshop on Languages and Compilers for Parallel Computing, 1994*, pp. 234–250 (Springer, Berlin, 1995). (ch. Flow-Insensitive Interprocedural Alias Analysis in the Presence of Pointers)
43. R.P. Wilson, M.S. Lam, Efficient context-sensitive pointer analysis for C programs, in *Proceedings of the International Conference on Programming Language Design and Implementation* (ACM, 1995), pp. 1–12
44. L.O. Andersen, Program Analysis and Specialization for the C Programming Language, Ph.D. dissertation (1994)

45. B. Steensgaard, Points-to analysis by type inference of programs with structures and unions, in *Proceedings of the International Conference on Compiler Construction* (1996), pp. 136–150
46. R. Ghiya, L. Hendren, Y. Zhu, Detecting parallelism in C programs with recursive data structures. IEEE Trans. Parallel Distrib. Syst. **1**, 35–47 (1998)
47. B.-C. Cheng, W.-M.W. Hwu, Modular interprocedural pointer analysis using access paths: design, implementation, and evaluation, in *Proceedings of the ACM SIGPLAN Conference on Programming Language Design and Implementation* (2000), pp. 57–69
48. J. Zhu, S. Calman, Symbolic pointer analysis revisited. SIGPLAN Notices **39**(6), 145–157 (2004)
49. B. Guo, N. Vachharajani, D.I. August, Shape analysis with inductive recursion synthesis. ACM SIGPLAN Notices **42**(6), 256 (2007)
50. L. Séméria, K. Sato, G. De Micheli, Resolution of dynamic memory allocation and pointers for the behavioral synthesis from C, in *Proceedings of the Design, Automation and Test Conference in Europe* (ACM, 2000), pp. 312–319
51. J. Babb, M. Rinard, A. Moritz, W. Lee, M. Frank, R. Barua, S. Amarasinghe, Parallelizing applications into silicon, in *Proceedings of the IEEE International Symposium on Field-Programmable Custom Computing Machines (FCCM)* (IEEE, New Jersey, 1999), pp. 70–80
52. P. O'Hearn, J. Reynolds, H. Yang, Local reasoning about programs that alter data structures, in *Computer Science Logic*, ed. by L. Fribourg, Lecture Notes Series, in Computer Science, vol. 2142, (Springer, Heidelberg, 2001), pp. 1–19
53. G. Winskel, *The Formal Semantics of Programming Languages: An Introduction* (MIT Press, Cambridge, MA, USA, 1993)
54. M. Raza, C. Calcagno, P. Gardner, Automatic parallelization with separation logic, in *Programming Languages and Systems* (2009), pp. 348–362
55. M. Botinčan, D. Distefano, M. Dodds, R. Grigore, M.J. Parkinson, coreStar: the core of jStar. Boogie 65–77 (2011)
56. C. Calcagno, D. Distefano, Infer: an automatic program verifier for memory safety of C programs, in *Proceedings of the International Conference on NASA Formal Methods* (Springer, Heidelberg, 2011), pp. 459–465
57. J. Berdine, C. Calcagno, P. O'Hearn, Symbolic execution with separation logic, in *Proceedings of the Asian Conference on Programming Languages and Systems (APLAS)* (2005), pp. 52–68
58. S. Magill, A. Nanevski, E. Clarke, P. Lee, Inferring invariants in separation logic for imperative list-processing programs, in *Proceedings of the Workshop on Semantics, Program Analysis, and Computing Environments for Memory Management (SPACE)* (2006), pp. 47–60
59. B. Cook, S. Magill, M. Raza, J. Simsa, S. Singh, *Making Fast Hardware with Separation Logic* (2010). http://www.cs.cmu.edu/~smagill/papers/fast-hardware.pdf
60. M. Botinčan, M. Dodds, S. Jagannathan, Proof-directed parallelization synthesis by separation logic. ACM Trans. Program. Lang. Syst. **35**(2), 1–60 (2013)
61. J. Villard, Here be wyverns! Verifying LLVM bitcode with llStar (2013). http://www.doc.ic.ac.uk/~jvillar1/pub/llstar-draft-oct13.pdf

# Chapter 4
# Heap Partitioning and Parallelisation

A crucial task in HLS of source code written in programming languages such as C/C++ is the extraction of parallelism from a sequential program description while preserving the program semantics. Additionally, parallelisation requires the memory system to match the computational parallelism. A fundamental difference of custom hardware implementations compared to microprocessors is the application-specific memory architecture. Instead of a monolithic memory space, the application data can be distributed over many small blocks of on-chip memory leading to a high aggregate memory bandwidth. Consequently, multiple computational units can be fed in parallel, which results in a very efficient parallelisation if expensive dynamic interconnects between any memory and any piece of computation can be eliminated, i.e. if the parallelism is communication-free. Automatic parallelisation for HLS compilers therefore requires a memory access and dependence analysis so as to detect parallelisation opportunities and partition the memory space accordingly. The objective in this chapter is to implement a static program analysis and automated code transformations that enable automatic parallelisation and distribution of data over separate blocks of on-chip memory.

Our program analysis and code transformations explicitly target programs that use pointers to heap-allocated data and dynamic memory allocation, a powerful and widely used feature of high-level programming languages such as C/C++. Automated program transformations that break the monolithic heap memory space into several portions (heaplets) and parallelise pointer-manipulating programs are beyond the scope of most current HLS techniques as we demonstrate in Chap. 2 and discuss in Chap. 3. This gap is mainly due to the difficulty of disambiguating pointer aliases and breaking the monolithic heap memory (implicit in the programming model) into small fragments. This thesis makes a step towards closing this gap and presents in this chapter a static analysis for pointer-manipulating programs which determines dependencies between loop iterations accessing heap memory and splits dynamic data structures into disjoint, independent regions. Our tool connects to the LLVM compiler infrastructure. The dependence/disjointness information provided by the


F. Winterstein, *Separation Logic for High-level Synthesis*, Springer Thesis,
DOI 10.1007/978-3-319-53222-6_4

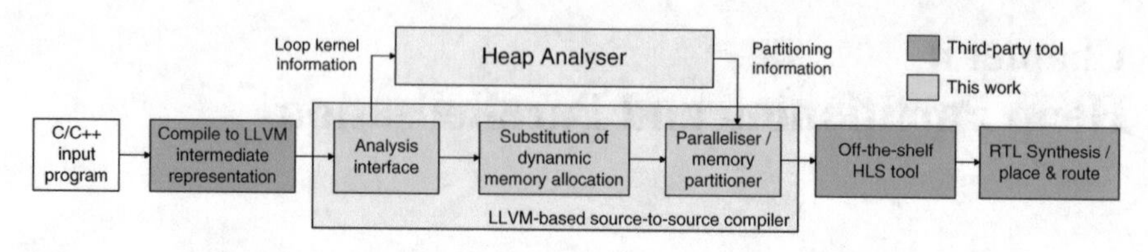


**Fig. 4.1** High-level compilation tool flow

analysis is passed to a source-to-source translator which modifies the code in such a way that a commercial-off-the-shelf HLS tool can parallelise the implementation and instantiate parallel memory blocks for the partitioned heap. Figure 4.1 summarises the high-level tool flow.

The main contribution of this work is the heap analyser in Fig. 4.1. The departure point from previous work is the use of recent advances in separation logic which extends classical logic by an operator that explicitly expresses the separation of resources, i.e. the non-aliasing property of two pointers. This paves the way for an automated program analysis and can straightforwardly handle dynamic memory allocation in disjoint heaps. The contributions of this chapter are:

- A separation logic-based parallelisation algorithm for pointer-manipulating programs that access dynamic data structures. Our static program analysis handles straight-line code as well as arbitrary `while`-loops and determines whether there is communication-free parallelism in the loop with respect to the accessed dynamic data structures. Starting from the C memory model of a global monolithic heap memory, it determines how to partition the heap and dynamic data structures into disjoint partitions that can be implemented in separate on-chip memory blocks (Sect. 4.2).
- The implementation of an automated source-to-source transformation infrastructure: The source translator ensures synthesisability of code containing unsupported constructs related to dynamic memory allocation (an unsupported feature in all common HLS tools). In a second pass, the disjointness information provided by our analysis is used to split the synthesised heap memory into separate blocks and to split a loop into multiple loops so as to obtain a semantically equivalent parallel implementation. The property of communication-free parallelism ensures that each functional unit only requires access to its own private memory block (Sect. 4.3).
- The demonstration of our tool flow using four real-life applications as test cases which build, traverse, update and dispose dynamically allocated data structures. The transformations at source code level allow us to stay as independent of a specific HLS tool as possible. We use Xilinx Vivado HLS as an exemplary back-end tool in our case studies. We also include hand-written HLS and RTL implementations for comparison (Sect. 4.4).

## 4.1 Motivating Example

Our running example, which we use throughout to illustrate the problem and our approach to solve it, is taken from the high-performance implementation of the tree-based $K$-means clustering algorithm discussed in Chap. 2. Listing 4.1 shows C-like pseudo code of the main kernel of the iterative filtering algorithm, the only difference from Chap. 2 being that the tree traversal here is destructive. Figure 4.2 shows the three heap-allocated data structures accessed by the loop: the tree, the centre sets, and the stack. The stack is implemented as a pointer-linked list whose head is modified by 'push' and 'pop' operations. The stack contains pointers to the tree nodes and centre sets. In Line 8, pointers to a centre set and tree node are fetched from the stack, and pointers to left and right child node as well as a newly allocated centre set (Line 13) are pushed onto the stack at the end of the loop body (Lines 16–17) - preceded by a data-dependent conditional (Line 15). The kd-tree is traversed in a pre-order fashion and visited nodes are deleted (Line 21).

```
1   //main kernel function
2   void filter(treeNode *root, centreSet cinit)
       {
3       centreSet* c0 = new centreSet;
4       *c0 = cinit;
5       stackRecord *s = push(root, c0, true,
          NULL);
6       while (s != NULL) {
7           treeNode *u; centreSet *c; bool d;
8           s = pop(&u, &c, &d, s);
9           centreSet cs = *c;
10          if (d) {
11              delete c;
12          }
13          centreSet *cnew = new centreSet;
14          *cnew = subfunction1(cs);
15          if  (u->left!=NULL) && (u->right!=
                NULL) && (subfunction2(cs)) {
16               s = push(u->left, cnew, true, s);
17               s = push(u->right, cnew, false, s
                   );
18          } else {
19              delete cnew;
20          }
21          delete u;
22      }
23  }
24
25  //auxiliary function push (create new list
       entry at head)
```

```
26    inline stackRecord* push(treeNode *u,
         centreSet *c, bool d, stackRecord *s){
27       stackRecord *t = new stackRecord;
28       t->u=u; t->c=c; t->d=d; t->n=s;
29       return t;
30    }
31
32    //auxiliary function pop (delete list head)
33    inline stackRecord* pop(treeNode **u,
         centreSet **c, bool *d, stackRecord *s){
34       *u=s->u; *c=s->c; *d=s->d; stackRecord *t
            =s->n;
35       delete s;
36       return t;
37    }
```

**Listing 4.1** C-like pseudo code of the (modified) main kernel of the filtering algorithm.

The static program analysis presented in Sect. 4.2 aims to determine the heap-carried data dependencies between loop iterations. Assuming that Fig. 4.2 describes the current state of the program, we can apply the following program transformations: (1) The remaining tree data structure (dark grey nodes) can be split into two sub-structures (two sub-trees labelled with *a*, one sub-tree labelled with *b*). (2) The linked list can be split into the uppermost node (pointing into the right sub-tree) and the nodes below (pointing into the left sub-tree). The same partitioning is applicable for the pool of centre sets. (3) The loop can be split into two loop kernels, each accessing one sub-tree, list segment and group of centre sets. The pointers dereferenced in any iteration of a loop will never access the data structures used by the other loop. Hence, once we have established that the loops are 'communication free' with respect to each other, we can split the heap memories into two banks of on-chip memory, each assigned to one loop as shown in Listing 4.2. A standard HLS tool can use the independence information to instantiate parallel hardware blocks for the loops without the need for arbitration of accesses to a global memory. Figure 4.3 shows the

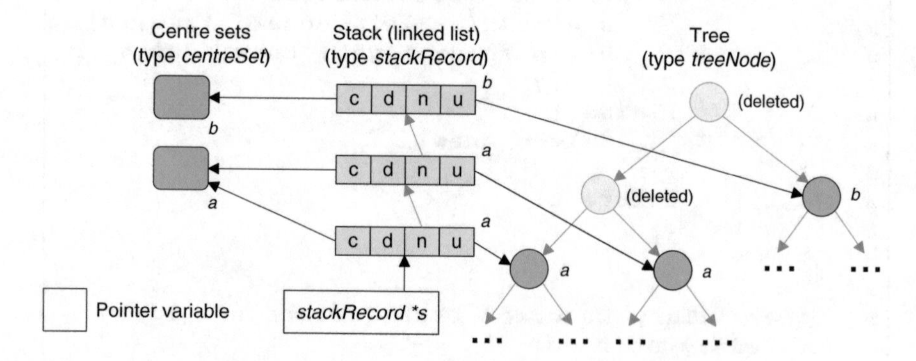


**Fig. 4.2** Snapshot of the linked data structures accessed by the loop in Listing 4.1

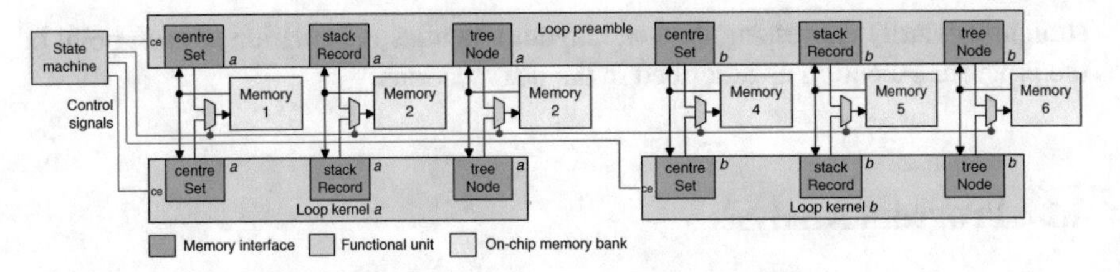


**Fig. 4.3** Synthesised hardware from the transformed code in Listing 4.2

hardware implementation that an HLS tool can synthesise from the modified code in Listing 4.2 using the memory partitioning and parallelisation information generated by our analysis. Without this analysis, state-of-the-art HLS tools, such as Vivado HLS, instantiate only one memory for the data types `centreSet`, `stackRecord` and `treeNode`, respectively. Furthermore, these tools are not able to detect that the loop kernels $a$ and $b$ can be scheduled for parallel execution as we have shown in Chap. 2.

```
1  //main kernel function
2  void filter(treeNode *root, centreSet
       cinit) {
3
4      ..preamble (pointers access
          partitions a and b)
5
6    while (s != NULL && s != s_b) {
7     // ... loop body (pointers access heap
          partition a only)
8    }
9
10   s = s_b;
11   while (s != NULL ) {
12    // ... loop body (pointers access heap
         partition b only)
13   }
14 }
```

**Listing 4.2** Transformed program from Listing 4.1 (two parallel loop kernels).

The difficult part of the above optimisation is the program analysis: regardless of scope, every two heap-directed pointers could potentially reference the same memory cell. The difficulty of analysing these programs increases with linked data structures which contain pointers in their link fields as discussed in Chap. 3. Ruling out aliasing requires an examination of the values that pointer variables hold during program execution. Separation logic addresses exactly this issue and provides a formalism for straightforwardly expressing the heap layout and alias information at each point of the program execution as described in the next sections.

## 4.2 Program Analysis

Our semantics-preserving parallelisation is based on the rationale that two program fragments can run in parallel if they access disjoint regions in memory (global variables being a special case of memory resources). We can then place each of these regions in physically separated on-chip memory banks without the need for cross-communication between functional units and each bank. Our memory partitioning and parallelisation analysis is hypothesis-based. The user specifies a value $P$. This value corresponds to the hypothesis that the heap accessed by the loop kernel can be split into $P$ disjoint parts and the loop can be split into $P$ parallel loops. The algorithm then tries to verify the hypothesis.

Proving the hypothesis is implemented in two main phases: searching for a necessary condition for the hypothesis to be true and, starting from the program state satisfying this condition, proving that the hypothesis is valid in all iterations. In the first phase, our tool symbolically executes the loop preamble and a finite number of loop iterations. During this process, it examines the separation logic formulae describing the accessed heap to determine whether the heap can be split into $P$ parts of identical shape, which is our necessary condition for partitioning. If such an initial partitioning can be established, the tool instruments the formulae with *cut-points* (markers) that mark the beginning of each partition. After the initial partitioning and instrumentation, the second phase is to prove that this partitioning is maintained not only in a finite number of iterations at loop start-up but in all loop iterations. Maintaining the partitioning in this case means that loop iterations (or parts of the loop body) are assigned to a heap partition and no iteration accesses the heap associated with a different partition than its 'own'. We use cut-points and heap footprint labels to assign heap partitions to loop iterations. Failing to prove the partitioning property in all iterations restarts the first phase. Generally, there are multiple options for the initial partitioning of the program state into $P$ portions. If the first option failed, the analysis tries the next one until we either obtain a successful proof or all options have been tested. Using the motivating example from Sect. 4.1, we first describe the initial partitioning and cut-point insertion followed by the proof of disjointness in all iterations.

A key building block of our analysis is the symbolic execution of a program. Section 3.5.3 introduced the general concept, Listing 5 gives a concrete example of the symbolic execution of the program fragment before the loop (loop preamble) in Listing 4.1 (Lines 3-5). For didactic reasons, we show pseudo code of this code section in the toy language defined in Definition 3.5 (Sect. 3.5.2, p. 46), interspersed with the separation logic formula (blue) describing the program state that is propagated from one statement to the next. The effect of each (atomic) command on the state formula is specified in Definition 3.6 (Sect. 3.5.2, p. 47) and the frame inference (Sect. 3.5.4,

**Listing 5** Symbolic execution of the loop preamble in Listing 4.1 (Lines 3-5).

$tree(root)$
1 : `new(c0);`
$tree(root) * c0 \mapsto \_$
2 : `[c0] := cinit;`
$tree(root) * c0 \mapsto cinit$
3 : `new(t);`
$tree(root) * c0 \mapsto cinit * t \mapsto [\mathtt{u}:\_, \mathtt{c}:\_, \mathtt{d}:\_, \mathtt{n}:\_]$
4 : `[t].u := root;`
$tree(root) * c0 \mapsto cinit * t \mapsto [\mathtt{u}: root, \mathtt{c}:\_, \mathtt{d}:\_, \mathtt{n}:\_]$
5 : `[t].c := c0;`
$tree(root) * c0 \mapsto cinit * t \mapsto [\mathtt{u}: root, \mathtt{c}: c0, \mathtt{d}:\_, \mathtt{n}:\_]$
6 : `[t].d := true;`
$tree(root) * c0 \mapsto cinit * t \mapsto [\mathtt{u}: root, \mathtt{c}: c0, \mathtt{d}: \mathtt{true}, \mathtt{n}:\_]$
7 : `[t].n := nil;`
$tree(root) * c0 \mapsto cinit * t \mapsto [\mathtt{u}: root, \mathtt{c}: c0, \mathtt{d}: \mathtt{true}, \mathtt{n}: \mathtt{nil}]$
8 : `s := t;`
$s = t \wedge tree(root) * c0 \mapsto cinit * t \mapsto [\mathtt{u}: root, \mathtt{c}: c0, \mathtt{d}: \mathtt{true}, \mathtt{n}: \mathtt{nil}]$
9 : ...

p. 49) ensures that only the part of the formula that is 'touched' by the command is updated. We furthermore assume that the sub-function `push` was inlined. The final state formula in Line 9 describes the pre-state of the `while`-loop, i.e. the state just before entering the loop body. Our analysis begins by instrumenting the loop pre-state as explained below.

### *4.2.1 Inserting Cut-Points*

Our analysis tries to split up spatial formulae at *cut-points*:

**Definition 4.1** (*Cut-point*) A cut-point is a program variable pointing to a heaplet in the program state formula.

The program can only interact with heap-allocated data via pointers (program variables). Useful heap partitioning requires the program to have access to each partition via pointers, e.g. given $ls(u, x_1') * ls(x_1', v) * ls(v, \mathtt{nil})$, the program can access the first and third list segment via cut-points $u$ and $v$, as opposed to the second list segment since $x_1'$ is not a cut-point (recall that a primed variable in a separation logic formula is not a program variable). The goal in this sub-section is to obtain $P$ cut-points in the pre-state of a loop iteration (i.e. the state before the loop body executes). This set of cut-points must satisfy certain conditions as we describe below. After the symbolic execution of the loop preamble as above, the program state is:

$$s = t \wedge tree(root) * c0 \mapsto cinit * t \mapsto [\mathtt{u}: root, \mathtt{c}: c0, \mathtt{d}: \mathtt{true}, \mathtt{n}: \mathtt{nil}] \quad (4.1)$$

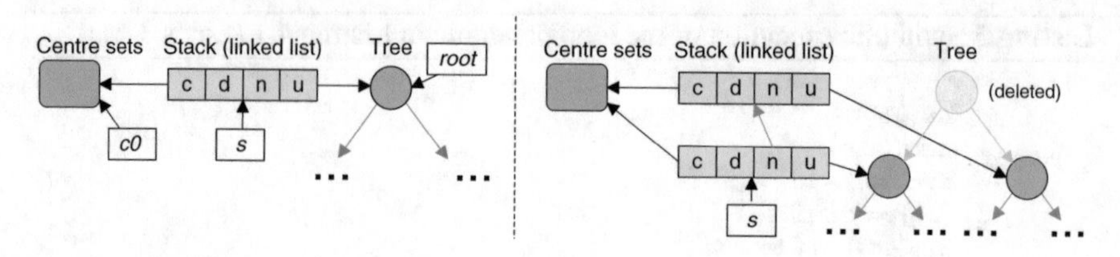


**Fig. 4.4** Pre-state before execution of the first (*left*) and the second loop iteration (*right*)

Figure 4.4, left, depicts (4.1), which contains the stack record (pointed to by $s$), the tree, and a centre set (pointed to by $c0$). Each heap predicate in (4.1) is also referenced by a cut-point. The cut-point insertion algorithm considers the program variable $s$ first and select the predicate $m_1 \equiv s \mapsto [\texttt{u}: root, \texttt{c}: c0, \texttt{n}: \texttt{true}, \texttt{n}: \texttt{nil}]$. Next, we try to find another predicate $m_2$ of the same shape as $m_1$ in the formula. To this end, we create a template $m_2 \equiv t'_0 \mapsto [\texttt{u}: t'_1, \texttt{c}: t'_2, \texttt{d}: t'_3, \texttt{n}: t'_4]$ and set $A \equiv (4.1)$. We then ask `coreStar`'s theorem prover whether it can match two predicates in $A$ with $m_1 * m_2$. If the prover is successful, $A$ contains the desired second predicate $m_2$ and we can extract it from the proof. If it is unsuccessful, we modify $A$ by symbolically executing the next iteration, which is the case in this example. The loop pre-state after 'peeling off' one loop iteration is (depicted in Fig. 4.4, right):

$$s = s'_2 \wedge tree(u'_1) * tree(u'_2) * c'_1 \mapsto \_ \tag{4.2}$$
$$* s'_2 \mapsto [\texttt{u}: u'_1, \texttt{c}: c'_1, \texttt{d}: \texttt{false}, \texttt{n}: s'_1] * s'_1 \mapsto [\texttt{u}: u'_2, \texttt{c}: c'_1, \texttt{d}: \texttt{true}, \texttt{n}: \texttt{nil}]$$

Now the matching is successful. We introduce a second cut-point $s_b$ and let it point to the only possible candidate $m_2$ by adding a conjunction to (4.2): $s_b = s'_1 \wedge (4.2)$. The new formula satisfies the necessary condition for partitioning: $s_b = s'_1 \wedge (4.2)$ contains $P = 2$ heaplets $m_1$ and $m_2$, of the same shape and referenced by cut-points. Next, we ask our proof engine described in the next section to prove that, in all subsequent loop iterations, the spatial part of the state can be split into $P = 2$ partitions, each of which being assigned either to cut-point $s$ or $s_b$. As explained in the next section, this proof fails here because of the lack of a second predicate $c_x \mapsto \_$ (the pointer aliasing is illustrated in Fig. 4.4, right). This means that this loop pre-state cannot be fully partitioned into two disjoint portions. Hence, we abandon the inserted cut-point, peel off another loop iteration, and reach the pre-state of the third iteration:

$$s = s'_4 \wedge tree(u'_3) * tree(u'_4) * tree(u'_2) \tag{4.3}$$
$$* c'_2 \mapsto \_ * c'_1 \mapsto \_ * s'_4 \mapsto [\texttt{u}: u'_3, \texttt{c}: c'_2, \texttt{d}: \texttt{false}, \texttt{n}: s'_3]$$
$$* s'_3 \mapsto [\texttt{u}: u'_4, \texttt{c}: c'_2, \texttt{d}: \texttt{true}, \texttt{n}: s'_1] * s'_1 \mapsto [\texttt{u}: u'_2, \texttt{c}: c'_1, \texttt{d}: \texttt{true}, \texttt{n}: \texttt{nil}]$$

The formula describes the program state shown in Fig. 4.2. We repeat the cut-point insertion as described above. Our tool explores all possible cut-point assignments (there are now multiple options now) and launches the proof engine in the next section for each candidate assignment. Assume we have assigned the second cut-point to the heaplet pointed to by $s_1'$: $s_b = s_1' \wedge (4.3)$. Starting from this pre-state, our proof engine can now successfully prove the parallelisation hypothesis of $P = 2$, because we have two valid cut-points $s = s_4'$ and $s_b = s_1'$, and the spatial part of the instrumented formula can be fully partitioned into two disjoint portions. These portions are independent in the sense that, starting from this pre-state, there is no subsequent loop iteration that accesses both portions. The next section explains how our analysis generates a proof of this fact. Note that, for other programs, we may not find a successful proof in which case we abort after $L_{\max}$ unrollings.

### 4.2.2 Proving Communication-Free Parallelism

The starting point for the proof engine is the program state obtained after the initial unrolling of a finite number of loop iterations above. In our example, we start with (4.3) and the two cut-points $s$ and $s_b$, and aim to split the heap accessed during the loop iterations into two portions $a$ and $b$. During symbolic execution of the loop body, we distinguish between two 'cut-point states' depending on whether we are currently accessing data structures 'belonging' to cut-point $s$ (portion $a$) or $s_b$ (portion $b$). Our tool constantly tracks the current cut-point state during symbolic execution of loop iterations. We switch to a different cut-point state once we have accessed a heaplet pointed to by a different cut-point variable as the one assigned to the current state. We assign label $a \in \mathrm{Lab}$ to all heaplets accessed during execution in cut-point state $a$ (cut-point $s$), and similarly for $b$ (cut-point $s_b$). We count *pointer dereferencing* and *delete* operations as an access. Our label assignment and cut-point state propagation through the program's CFG are implemented as add-ons to `coreStar`. Tracking the cut-point state together with footprint label assignment to heaplets allows the analysis to assign heap partitions to loop iterations.

The parallelisation goal is to partition the loop iteration space into two groups labelled $a$ and $b$, and we try to establish the fact that a heaplet accessed by an iteration in cut-point state $a$ (of group $a$) is never accessed by another iteration of group $b$. In other words, we try to prove that the separation of the accessed heap into $a$ and $b$ is invariant in each subsequent loop iteration. If the number of iterations was known at compile time, we could symbolically execute all iterations to prove this property. However, in general, this number is not statically determinable because of the data dependent loop condition (Listing 4.1, Line 15). Hence, we perform a *fix-point calculation* [1, 2] for proving that the separation property is loop invariant.

The fix-point calculation performs a symbolic execution of a sequence of loop iterations. Doing so, it aims to find a generalised formula that describes the program state in *all* loop iterations. Once such as formula has been constructed, we say the fix-point calculation has *converged* and terminates. The fix-point calculation consists of two main components: (1) rewriting the state formula in the quest for a generalised state representation while it symbolically executes loop iterations, and (2) deciding when the fix-point iteration has converged. The former part is based on abstracting the current state formula, which folds singleton heaplets in recursive predicates such as those in Definitions 3.2–3.4 (p. 46) using the abstraction rules described in Sect. 3.5.4 (pp. 49–51). In line with [1], our abstraction rules forbid folding across program variables. Note that this also prevents folding across cut-points. The abstraction step prevents accumulating singleton heaplets such as $s \mapsto [\mathtt{n} : x_1']$ during repeated execution of the loop body and is crucial for convergence of the fix-point calculation. Our fix-point calculation adopts and modifies the technique described by Magill et al. [1] and works as follows:

1. Start with the pre-state of the loop $M_0^{pre}$ equal to (4.3) with cut-points $s$ and $s_b$ inserted.
2. Symbolically execute $\{M_i^{pre} \wedge b\}$ '`loop body`' $\{M_{i+1}^{post}\}$, $b$ is the loop condition, $i$ is the iteration counter and $M_{i+1}^{post}$ describes the program state after the loop body in iteration $i$ has been executed. We attach labels $a$ or $b$ to heaplets corresponding to the current cut-point state. If we find both labels $a$ and $b$ on a heaplet, it means that this heaplet has been accessed by at least one iteration of cut-point state $a$ and one of state $b$; the separation into disjoint partitions is not maintained and we abort, report a failed proof and restart the cut-point insertion to obtain a different initial partitioning. If only either $a$ or $b$ are attached to any heaplet we continue with the next step.
3. Absorb singleton heaplets in $M_{i+1}^{post}$ in recursive predicates such as those in Definitions 3.2–3.4 (Sect. 3.5.1) using the abstraction rules defined Sect. 3.5.4. This results in a rewritten form of $M_{i+1}^{post}$ if an abstraction rule can be applied.
4. The fix-point calculation terminates if $M_{i+1}^{post}$ implies a post-state of one of the previous iterations $M_k^{post}$, $k = 0, ..., i$. Formally, we ask `coreStar`'s theorem prover to decide $M_{i+1}^{post} \vdash \bigvee_{k=0..i} M_k^{post}$ (the right hand side is the disjunction of all previous post-states). If the implication could not be shown to hold we set $M_{i+1}^{pre} := M_{i+1}^{post}$ and continue with step 2).

For our example, we reach a fix-point after 7 iterations of steps 1) to 4). Note that, for another candidate for the cut-point assignment in (4.3) ($s_b = s_3'$ instead of $s_b = s_1'$) as discussed above, the fix-point calculation would have been aborted because we had eventually reached the state $\langle c_2' \mapsto \_\rangle_{\{a,b\}}$ (the label set $\{a, b\}$ denotes sharing between functional unit $a$ and $b$).

The successful fix-point calculation tells us that the heap accessed by the loop, after peeling off a finite number of initial loop iterations, can be partitioned into two disjoint regions labelled $a$ and $b$. Furthermore, it tells us that the partitioning will be maintained for all following loop iterations, each of which will either access heap portion $a$ or $b$, but not both. A code transformation can now split the original code into two code fragments, each having access to its own heap partition as shown in Listing 4.2. What remains is to assign all heap-manipulating program statements in the loop preamble and initially unrolled iterations to the correct partitions. This is described in the following section.

### 4.2.3 Assigning Heap Partition Information to Statements

After the analysis has determined that the loop can be split into two loops with access to their private heap partitions, we must ensure that the pointers used in the preamble and unrolled iterations refer to the correct memory partition. For example, the predicate $s'_4 \mapsto [\texttt{u} : u'_3, \texttt{c} : c'_2, \texttt{d} : \texttt{false}, \texttt{n} : s'_3]$ in (4.3) obtains the partition label $a$ during the loop analysis: $\langle s'_4 \mapsto [\texttt{u} : u'_3, \texttt{c} : c'_2, \texttt{d} : \texttt{false}, \texttt{n} : s'_3]\rangle_{\{a\}}$. The heaplet described by this predicate, however, was allocated (new statement, Listing 4.1, Line 17) and written to (pointer dereferencing, also Line 17) in the second iteration that was peeled off during the cut-point insertion. Consequently, we must attach the partition information to these program statements as well.

We link the partition assignment to heap-manipulating program commands with a combination of our labelled symbolic execution (footprint labels according to the cut-point state) with the standard labelled symbolic execution in [3] (a unique footprint label for each program statement). Recall that (4.3) describes the program state just before launching the fix-point calculation. During the fix-point calculation, we record each heaplet the first time it gets assigned a label. Recording on first label assignment is necessary because, for instance, we may lose track of the predicate $c'_2 \mapsto \_$ in (4.3) as it will be disposed (Listing 4.1, Line 11) during the course of fix-point calculation before we even access $c'_1 \mapsto \_$ for the first time. After a successful fix-point calculation, we stitch together all snapshots, resulting in a labelled version of (4.3):

$$\begin{aligned} s = s'_4 \wedge s_b = s'_1 \wedge \langle tree(u'_3)\rangle_{\{a\}} * \langle tree(u'_4)\rangle_{\{a\}} * \langle tree(u'_2)\rangle_{\{b\}} \\ * \langle c'_2 \mapsto \_\rangle_{\{a\}} * \langle c'_1 \mapsto \_\rangle_{\{b\}} * \langle s'_4 \mapsto [\texttt{u} : u'_3, \texttt{c} : c'_2, \texttt{d} : \texttt{false}, \texttt{n} : s'_3]\rangle_{\{a\}} \\ * \langle s'_3 \mapsto [\texttt{u} : u'_4, \texttt{c} : c'_2, \texttt{d} : \texttt{true}, \texttt{n} : s'_1]\rangle_{\{a\}} * \langle s'_1 \mapsto [\texttt{u} : u'_2, \texttt{c} : c'_1, \texttt{d} : \texttt{true}, \texttt{n} : \text{nil}]\rangle_{\{b\}} \end{aligned} \tag{4.4}$$

During the symbolic execution of the loop preamble and iteration unrolling prior to the fix-point calculation, we also record the program statements that accessed each of the heaplets in (4.4) by assigning a second set of footprint labels ($FT$) as in the standard label assignment in [3]. This set contains a unique label for each accessing statement, e.g. $FT = \{l_2, l_3, l_7\}$ for statements 2, 3 and 7. With these two label sets we obtain a mapping

$$m : \texttt{Lab} \rightarrow \{a, b\} \tag{4.5}$$

where Lab is the set of all unique labels assigned to heap-manipulating program commands in the loop preamble and unrolled iterations. This mapping allows us to assign the correct heap partition information to each pointer access. This information is used by the source-to-source transformation for correct code instrumentation.

The above analysis provides both memory partitioning information (by labels assigned to heaplets) and the legality of parallelisation (by a successful fix-point calculation). Algorithm 6 summarises our heap analysis. The heap analysis in this section focuses on the function under test (filter in our motivating example) in isolation. However, some data structures used by the function under test may have been built up by different parts of the program that are external to this function. We

**Algorithm 6** Heap partitioning analysis

```
 1: Input:
 2: loop body specification (code)
 3: initial state formula (Π ∧ Σ{FT})^initial (from symbolic execution of loop preamble)
 4: parallelisation hypothesis P
 5: Output:
 6: validity of parallelisation hypothesis (success)
 7: number of initial unrollings required (it)
 8: label mapping: program statement identifiers to heap partitions (m)
 9: Variables:
10: it                      ▷ Iteration counter (number of iterations to be unrolled)
11: C                       ▷ set of cut-points
12: S_cutpoints             ▷ set of cut-point states
13: Π ∧ Σ{FT}               ▷ state formula in separation logic (attached footprint label set FT)
14: Π ∧ Σ{CS}               ▷ state formula in separation logic (attached cut-point state set CS)
15: m                       ▷ label mapping m: FT → CS

16: function HEAP- PARTITIONING
17:     it ← 0
18:     C ← ∅
19:     Π ∧ Σ{FT} ← (Π ∧ Σ{FT})^initial
20:     success ← false
21:     repeat
22:         while not checkIfValidCutpInsertion(Π ∧ Σ{FT}, C) do
23:             Π ∧ Σ{FT} ← SymbExec(Π ∧ Σ{FT}, it)        ▷ peel off it iterations (Sect. 4.2.1)
24:             Π ∧ Σ{FT}, C ← CutpInsert(Π ∧ Σ{FT}, P)    ▷ insert P cut-points (Sect. 4.2.1)
25:             it ← it + 1
26:         end while
27:         S_cutpoints ← AssignCPStates(C)              ▷ assign states to cut-points (Sect. 4.2.2)
28:         Π ∧ Σ{CS}, success ← FixpCalc(Π ∧ Σ{FT}, C, S_cutpoints)   ▷ fix-point calculation
    (Sect. 4.2.2)
29:         m ← GetLabelMapping(Π ∧ Σ{CS}, Π ∧ Σ{FT})    ▷ label mapping (Sect. 4.2.3)
30:     until success or it ≥ L_max
31: end function
32: return success, it, m
```

call such data structures the *context* of the function under test. Our loop analysis generates partitioning information for all heap-allocated data structures it uses. In some cases, after the heap partitioning by the loop analysis above, we may wish to transfer the partitioning information to the enclosing program. In the appendix A, we describe an approach that can be used to extend our technique above to a context-aware analysis.

The next section explains how the partitioning information and the legality of parallelisation are used in a source-to-source translator for automated code optimisation.

## 4.3 Implementation

Our tool flow consists of three main parts: the heap analyser, a source-to-source compiler, and a set of third-party tools (back-end HLS and RTL synthesis tools). Figure 4.5 shows the complete tool flow.

### 4.3.1 Heap Analyser

Our heap analyser connects to the analysis interface of the source translator and implements the two-step analysis described above. It is written in OCaml and is based on our modified version of `coreStar`. We we extended `coreStar` to include labelled symbolic execution and cut-point processing and modified it to generate the disjointness proofs based on non-overlapping footprint label sets as described in Sect. 4.2.2. Our heap analyser currently uses 122 logic rules as described in Sect. 3.5 which define pure and spatial predicates, such as those in Definitions 3.2–3.4 (p. 42), and how footprint labels are propagated. These rules also define, for example, under what conditions a points-to predicate describing a singleton list node can be 'gobbled up' by an existing linked list predicate in order to ensure convergence of the fix-point calculation as described in Sect. 4.2.2.

The symbolic execution is performed on the control flow graph of the program which is built internally by `coreStar`. It operates on a representation of the input program in the *coreStar intermediate language* (`coreStarIL`) [4]. This language consists of Hoare triple-like statements which specify the effect of program commands on the program state (using separation logic specifications). It also contains constructs for control flow in order to implement branching and loops. Real-world input code must be translated into `coreStarIL` before the program can be analysed. Translating C/C++ code directly into `coreStarIL` is a complex task. Our approach is to first compile the C/C++ input into the LLVM IR using third-party front-ends and then derive a specification in `coreStarIL` from the LLVM code. We choose the LLVM IR as the input language to our tool because many state-of-the-art HLS tools [5–9] compile the input code into the LLVM IR prior to RTL generation. This choice greatly improves the level of automation and paves the way for the integration of our technique into existing HLS flows.

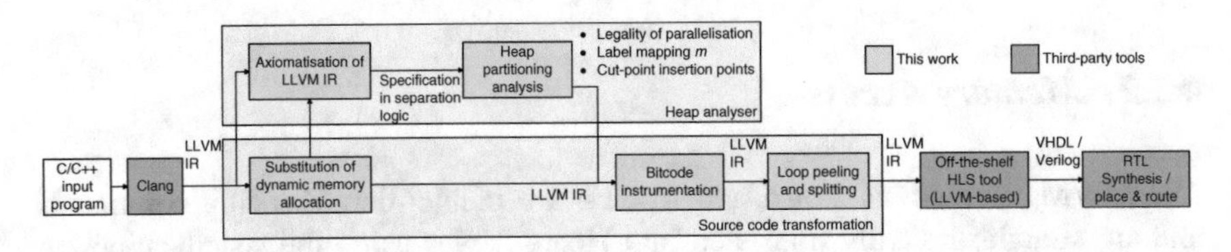


**Fig. 4.5** LLVM-based CAD flow including the heap analyser, source translator and third party tools for HLS and RTL implementation

The LLVM IR code style in this thesis is that of the IR generated by `Clang` [10] from C++ code. LLVM IR is a typed assembly-like language with an unbounded number of virtual registers [11]. The types are arbitrary-width integers (`i`$N$ for $N$ bits), standard floating-point types (`half`, `float`, `double`), pointers, function and array types, and structures (`struct` in C). We denote a type as $t$ and $t*$ is the corresponding pointer type. The type is always attached to a value that appears in an LLVM instruction. We denote a value as $v$. Instructions are grouped into *basic blocks*. Each basic block has a label (denoted by $l$). For brevity, we do not discuss the whole LLVM instruction set, but focus on a subset:

$$
\begin{aligned}
I ::= {} & x\ \texttt{=}\ \texttt{load}\ t*\ v \\
| {} & \texttt{store}\ t\ v_1\texttt{,}\ t*\ v_2 \\
| {} & x\ \texttt{=}\ \texttt{icmp}\ cond\texttt{,}\ t\ v_1\texttt{,}\ v_2 \\
| {} & \texttt{br i1}\ v\texttt{,}\ \texttt{label}\ l_{\mathrm{T}}\texttt{,}\ \texttt{label}\ l_{\mathrm{F}} \\
| {} & \texttt{br label}\ l \\
| {} & x\ \texttt{=}\ \texttt{getelementptr}\ t\ v*\texttt{,}\ t_1\ v_1\texttt{,}\ \texttt{...,}\ t_n\ v_n \\
| {} & x\ \texttt{=}\ \texttt{phi}\ t\ \texttt{[}v_1\texttt{,}\ l_1\texttt{],}\ \texttt{...,}\ \texttt{[}v_n\texttt{,}\ l_n\texttt{]} \\
| {} & x\ \texttt{=}\ \texttt{call}\ t\ f\texttt{(}t_1\ v_1\texttt{,}\ \texttt{...,}\ t_n\ v_n\texttt{)}
\end{aligned}
$$

The `load` instruction dereferences the pointer $v$ and loads the memory content into the variable (register) $x$. Similarly, `store` dereferences $v_2$ for a write access. The `icmp` instruction checks if $v_1$ and $v_2$ satisfy the condition *cond* and returns a boolean value (type `i1`). This result can be used by the conditional branch instruction which directs the control flow to the basic blocks $l_{\mathrm{T}}$ (true) or $l_{\mathrm{F}}$ (false), respectively, based on the value $v$. The `getelementptr` instruction returns a pointer to an element inside a data structure (array or structure). The returned address is calculated by adding several offsets $(v_1, \ldots, v_n)$ to the base pointer $v$. For example, the pointer to the 4th field of a structure called `S` is returned by `getelementptr %struct.S* %y, i32 0, i32 4`, where `%y` is the base pointer (the first offset is set to `0` by default if a field in a singleton structure is referenced). A `phi` node is a standard component of a program in the single static assignment form. The variable $x$ is assigned one of the values $v_1, \ldots, v_n$ depending on from which of the basic blocks $l_1, \ldots, l_n$ the control flow arrives at the current block.

The following sections describe how we translate the LLVM instructions above into axiomatic specifications in separation logic, expressed in `coreStarIL`.

### 4.3.2 Memory Access

The LLVM instructions `load` and `store` are pointer-dereferencing commands and are straightforwardly translated into Hoare triples using the specifications in Definition 3.6 (Sect. 3.5.2, p. 47). Access to fields in structures is implemented as a pair of a `getelementptr` and a `load`/`store` instruction and is specified in the same way. For example, a write access to the $i$-th field of a structure ($1 < i < n$) is a `getelementptr/store` pair:

$$x = \texttt{getelemtptr}\ t{*}\ v,\ \texttt{i32 0, i32 i;}\quad \texttt{store}\ t_i\ v_i,\ t_i{*}\ x$$

Our source code processor looks for such `getelementptr-load/store` patterns creates a pointer-dereferencing specification for them.

### 4.3.3 *Dynamic Memory Allocation*

Heap allocation and deallocation (`new/delete`) are implemented with calls to standard library functions in LLVM: `@_Znwj` / `@_Znwm` for `new` and `@_ZdlPv` for `delete`. Calls to library functions for `new` and `delete` are specified at the call site as described in Definition 3.6 with the *emp* predicate in the pre- or post-condition, respectively.

### 4.3.4 *Control and Data Flow*

Branching and loops are implemented with `br` instructions in LLVM. Branch instructions target labelled basic blocks and each block is usually terminated by a `br` instruction. Our symbolic execution engine constructs a control flow graph of the program using the same concept of labelled blocks and unconditional `goto` statements of the `coreStarIL`. Unconditional `br` instructions straightforwardly translate into `goto` statements:

$$\texttt{br label}\ l \rightsquigarrow \texttt{goto}\ l$$

The `goto` statement causes our symbolic execution to jump and process the Hoare triples below the label $l$. If multiple labels are given to `goto`, the analysis explores each control flow path. Conditional branching in LLVM IR consists of an `icmp`[1]-`br` pair. We translate it as follows:

$$\begin{array}{lcl}
x = \texttt{icmp}\ \mathit{cond},\ t\ v_1,\ v_2; & & \texttt{goto}\ l_{\mathrm{T}},\ l_{\mathrm{F}};\\
\texttt{br i1}\ x,\ \texttt{label}\ l_{\mathrm{T}},\ \texttt{label}\ l_{\mathrm{F}}; & & l_{\mathrm{T}}:\\
l_{\mathrm{T}}: & & \{\}\ .\ \{\ \mathit{cond}(v_1, v_2)\ \}\\
\ldots & \rightsquigarrow & \ldots\\
l_{\mathrm{F}}: & & l_{\mathrm{F}}:\\
\ldots & & \{\}\ .\ \{\ \neg\mathit{cond}(v_1, v_2)\ \}\\
 & & \ldots
\end{array}$$

[1] We omit `fcmp` for floating-point comparisons here.

The expression '{ } . { $cond(v_1, v_2)$ }' means that the analysis adds the condition $cond(v_1, v_2)$ to the control flow path of $l_T$. This is equivalent to an `assume` statement. The condition encodes (dis)equality, less or greater relations between $v_1$ and $v_2$ ($\neg cond(v_1, v_2)$ is the negated condition). In general, $v_1$ and $v_2$ are symbolic values and the analysis cannot decide in which of the two control flow paths the condition is satisfied: it explores both paths in this case. However, in some cases, some information about the values of $v_1$ and $v_2$ may be available in the current state formula and the analysis may be able to terminate the path that is inconsistent with the branch condition.

If a variable is assigned different values based on the flow of control, the LLVM IR uses `phi` nodes. We show an example of a specification of the `phi` node with two source blocks, $l_1$ and $l_2$, below. Whether the variable $x$ receives the value $v_1$ or $v_2$ depends on the block from which the jump to $l_0$ is made.

| | | |
|---|---|---|
| $l_0$: | | $l_0$: |
| $x =$ `phi` $t$ $[v_1, l_1]$, $[v_2, l_2]$ | | ... |
| ... | | $l_1$: |
| $l_1$: | | ... |
| ... | | $\{x = y_1'\}$ . $\{ x = v_1 \}$ |
| `br label` $l_0$ | $\rightsquigarrow$ | `goto` $l_0$ |
| $l_2$: | | $l_2$: |
| ... | | ... |
| `br label` $l_0$ | | $\{x = y_1'\}$ . $\{ x = v_2 \}$ |
| | | `goto` $l_0$ |

We treat `phi` nodes in a similar way as conditional branches in that we add artificial assignment specifications in the source blocks. An assignment is represented by the triple '$\{x = y_1'\}$ . $\{ x = v_1 \}$'. The assignments are placed at the bottom of the blocks just before the terminator instruction.

The above definitions show how we represent a heap-manipulating program in LLVM IR with Hoare triples using separation logic formulae. This representation is directly encoded in a `coreStarIL` representation of the input program and can thus be processed by the symbolic execution engine in our version of `coreStar` in order to implement the heap analysis in Sect. 4.2. The output of this analysis is a flag indicating the successful heap partitioning (*success*), the number of initially unrolled iterations (*it*), and a table assigning program statement identifiers to heap partitions (*m*). The next section describes how our source-to-source transformation uses this information.

### 4.3.5 *Source-to-Source Compiler*

A previous version [12] of our source translator was built on the C-based ROSE source compiler infrastructure [13]. In a later refinement of our tool, we moved the source-to-source transformations entirely to LLVM IR in order to ensure a tighter integration with the program analysis and to canonicalise the code transformations. Our code transformation works on the LLVM IR generated from C/C++ code using the `Clang` front-end and is implemented as a custom LLVM pass [11].

After parsing in the LLVM code, the source transformation first replaces the basic routines for dynamic memory allocation with custom implementations to ensure synthesisability by an off-the-shelf HLS tool. The heap is replaced by arrays, which will be synthesised into on-chip block RAM by the HLS tool, and the corresponding pointers are converted to integer variables (`i32`). Occurrences of `new` and `delete` operations are grouped according to the type of their operand and custom allocator functions are instantiated for each type as a replacement. Dynamic type casts are currently not supported. Our fixed-size allocator is a standard implementation using a *free-list* which keeps track of occupied memory space. It is implemented in a file which contains template LLVM functions for dereferencing, allocation and disposal and which is automatically included by our tool. We stress that this work focuses on memory partitioning and parallelisation and is therefore orthogonal to work that determines a bound on the amount of allocated heap memory. Cook et al. [14] describe a technique for finding parametric worst-case bounds on the heap consumption based on a separation logic-driven analysis which could be used for this purpose in our benchmarks. However, as we shall see in the next chapter, we approach this issue in a different way by moving the heap memory space into off-chip DRAM and host system main memory.

```
1  // t->u = u; original LLVM code
2  %1 = getelementptr inbounds %struct.stackRecord* %t, i32 0, i32 0
3  store %struct.treeNode* %u, %struct.treeNode** %1
```

**Listing 4.3** Original LLVM IR of the statement `t->u = u`.

```
1  // t->u = u; transformed LLVM code
2  %_aux1 = call %struct.stackRecord* @auxMakePointer_0(%struct.stackRecord*
       getelementptr inbounds ([65536 x %struct.stackRecord]* @heap_partition0, i32
       0, i32 0), i32 %t)
3  %1 = getelementptr inbounds %struct.stackRecord* %_aux1, i32 0, i32 0
4  store i32 %u, i32* %1
```

**Listing 4.4** Transformed LLVM IR (dereferencing in heap partition 0).

In the last step of the transformation, the memory partitioner/paralleliser receives information from the heap analyser that a parallelisation is legal and how the heap arrays have to be partitioned. The heap partition information is passed to the code transforming pass via LLVM *metadata*, additional information that can be attached to an LLVM instruction. The arrays representing the heap memory are partitioned according to the metadata information. Dereferencing of heap-directed pointers is substituted using an auxiliary pointer variable indexing the heap array. Listings 4.3 and 4.4 show an example for the dereferencing `t->u = u` in heap partition 0. The auxiliary pointer variable in this case is `_aux1`. The calls to (de-) allocation functions must be customised similarly: the scope of `new/delete` operations is restricted to its heap array partition and we instantiate an allocator, including the free-list, for each partition.

The parallelisation analysis, if successful, has divided the loop iterations into $P$ independent groups, where $P$ is the degree of parallelisation. Additionally, several loop iterations may have been peeled off by the analysis as it is the case in our motivating example described above. Our source transformation removes the original loop and inserts two sections of code: (1) The original loop body guarded by an `if` conditional with the loop condition representing the iterations that have been unrolled during the analysis. (2) $P$ loops of the same type and with the same loop condition as the original one, each containing the fragment of the loop body which accesses one of the independent groups. We must also ensure that the cut-point insertion is reflected in this code transformation. In (4.4), we added the additional conjunction forcing $s_b = s_1'$, which means that the code transformer must add an assignment instruction to the new variable $s_b$ somewhere in the loop preamble or the unrolled iteration. We obtain the information as to where to insert this instruction from the set of instruction identifier labels above which is attached to each heaplet. From (4.4), we can easily find the heaplet referenced by $s_b$ (in this case $s_1' \mapsto [\mathtt{u} : u_2', \mathtt{c} : c_1', \mathtt{d} : \mathtt{true}, \mathtt{n} : \text{nil}]$). We take the first identifier label from the label set attached to this heaplet and obtain the LLVM instruction after which the assignment should be placed. The last step is to extract the pointer operand of this instruction and add metadata information that $s_b$ must be assigned this value after the instruction. The bitcode instrumentation is responsible for embedding the heap partition and cut-point insertion data in the LLVM metadata.

The LLVM IR of the input code is finally restructured in a way that exposes parallelism and ensures the correct assignment of heap partitions to parallel on-chip memory banks. The generated LLVM IR is then passed to a down-stream HLS tool as shown in Fig. 4.5. The next section describes the evaluation of our memory partitioning and loop parallelisation tool flow on four pointer-chasing benchmarks applications. It also give insights into the analysis complexity and tool run-time.

## 4.4 Experiments

We test the tool flow in Fig. 4.5 using C++ implementations taken from real-world applications. We use Xilinx Vivado HLS 2014.4 as a back-end HLS tool and Xilinx Vivado 2014.4 for RTL synthesis. However, since our optimisations are at source code level, our tool can be also used in combination with a different HLS tool. Our benchmark applications are:

**Merger**. The program maintains four linked lists whose nodes are sorted according to a key. It repetitively reads four key-value pairs from its interface and performs a sorted insertion in each list for each pair. After a constant number of pairs has been received, it repeatedly deletes the head node of that list which contains the smallest key until all lists are empty. The output is a sorted sequence of all key-value pairs. A distinguishing feature of this applications is that the loop under analysis contains a sub-loop. During each symbolic execution of an outer loop iteration the proof engine requires a few inner iterations to converge to a loop invariant for the inner sub-loop. We consider this benchmark a representative example from the class of list processing programs.

**Tree Deletion**. This application performs a full traversal of a pointer-linked tree data structure and deletes the visited tree nodes after some computation using the node data.

**Filter**. This is the motivating example in Sect. 4.1 which is taken from the direct implementation of the filtering algorithm for efficient $K$-means clustering [15]. Our tool splits the loop in Listing 4.1 and partitions the heap memory with degree $P$. The code fragment is embedded in a larger program which includes tree build-up and centre processing to form a complete clustering application. This example is interesting in that it is more complicated than a usual toy example: loop iterations allocate and dispose centre sets, preceded by a data-dependent conditional, which carry a heap dependence between some iterations. Our analysis detects that there are no heap-carried dependencies between iterations which access tree nodes without a parent-child relation.

**Reflect Tree**. The application traverses a binary tree in pre-order fashion and recursively swaps the left and right child pointer of each node, thus producing a mirrored tree. It also performs some computation at each node and updates the data fields of the tree nodes.

The target device is a Virtex 7 FPGA (Xilinx VC707 evaluation board, xc7vx485 tffg1761-2) and all results are taken from placed and routed designs. We report resource utilisation in LUTs, FFs, DSP slices and 36k-BRAMs. We also report the achieved clock speed (target 200 MHz) and the time required for task completion (latency) which we derive from the achieved clock rate and the clock cycle count determined via simulations of the generated RTL designs. The RTL test benches for the benchmarks are fed with application-specific input data. For each test case, Table 4.1 shows the implementation results for three cases: The *baseline* case shows the implementation if the tool only ensures synthesisability (syntactical substitution of dynamic memory allocation and heap-directed pointers, no heap analysis) without parallelisation. The second case shows the results of ‘blind’ loop unrolling. Instead of using our source-to-source compiler, we use the standard Vivado directive for partial loop unrolling here which instantiates $P$ parallel loop kernels. We call this case ‘blind parallelisation’ because it is not guided by our heap analysis and no heap partitioning is performed by Vivado HLS. The third row shows results if the tool flow uses the heap analyser for memory partitioning and parallelisation using our source transformation (automatic parallelisation with degree $P$), an optimisation that cannot be done by Vivado HLS itself as shown in the previous case and as explained in Chap. 2. The loop peeling factor $PF$ indicates the number of loop iterations that were peeled off during the cut-point insertion as described on Sect. 4.2.1. The speed-up $S$ relates the latency of the automatically parallelised benchmarks to that of the base line case.

**Table 4.1** Implementation results and comparison

*P*: parallelisation degree; *PF*: peeling factor (number of initially unrolled iterations); *S*: speed-up over baseline

| | *P* | *PF* | LUT | FF | DSP | BRAM | Clock rate | Latency | *S* |
|---|---|---|---|---|---|---|---|---|---|
| **Merger** (4 × 2048 random input key-value pairs) | | | | | | | | | |
| Base line (reference) | 1 | 0 | 1644 | 1547 | 0 | 96 | 143 MHz | 88.8 ms | 1.00 |
| Blind unrolling | 4 | 0 | 1969 | 1957 | 0 | 96 | 141 MHz | 90.4 ms | 0.98 |
| Autom. parallelisation **(this work)** | 4 | 0 | 2012 | 1901 | 0 | 82.5 | 198 MHz | 16.7 ms | **5.31** |
| **Tree Deletion** (16383 tree nodes) | | | | | | | | | |
| Base line (reference) | 1 | 0 | 3016 | 4139 | 9 | 515 | 193 MHz | 9827.2 us | 1.00 |
| Blind unrolling | 2 | 0 | 3818 | 5478 | 12 | 515 | 190 MHz | 9798.4 us | 1.00 |
| Autom. parallelisation **(this work)** | 2 | 1 | 6802 | 10508 | 27 | 515 | 192 MHz | 5353.3 us | **1.84** |
| **Filter** (16384 3-dimensional data points, 32767 tree nodes, $K = 128$ clusters) | | | | | | | | | |
| Base line (reference) | 1 | 0 | 8387 | 4981 | 18 | 609.5 | 181 MHz | 5390.6 us | 1.00 |
| Blind unrolling | 2 | 0 | 9746 | 6832 | 36 | 609.5 | 179 MHz | 5718.0 us | 0.94 |
| Autom. parallelisation **(this work)** | 2 | 2 | 14197 | 12145 | 72 | 614.5 | 200 MHz | 2860.3 us | **1.88** |

(continued)

**Table 4.1** (continued)

*P*: parallelisation degree; *PF*: peeling factor (number of initially unrolled iterations); *S*: speed-up over baseline

| | *P* | *PF* | LUT | FF | DSP | BRAM | Clock rate | Latency | *S* |
|---|---|---|---|---|---|---|---|---|---|
| **Reflect Tree** (16383 tree nodes) | | | | | | | | | |
| Base line (reference) | 1 | 0 | 1942 | 2576 | 12 | 291 | 200 MHz | 3768.2 us | 1.00 |
| Blind unrolling | 2 | 0 | 2191 | 2987 | 21 | 291 | 200 MHz | 3809.4 us | 0.99 |
| Autom. parallelisation (**this work**) | 2 | 1 | 4256 | 6472 | 36 | 291 | 200 MHz | 2037.0 us | **1.85** |

Vivado HLS is unable to parallelise any of the benchmarks in the blind unrolling case, i.e. without explicit heap partitioning. Including a directive for implementing dual-port memories to increase the number of access ports did not have any influence on the scheduling in our cases. Blind unrolling consumes more resources at the same execution time as the baseline. On the other hand, our heap analysis detects the independence of the four linked lists in the **Merger** benchmark and parallelises the application. The speed-up in terms of cycle count is close to the maximum speed-up of $P = 4$ and the automatically memory partitioned design achieves a higher clock rate than the baseline and blind unrolling case, resulting in a run-time advantage of 5.31× over the base case. The analysis also partitions the data structures of **Filter**, **Tree Deletion** and **Reflect Tree** which enables successful parallelisation (speed-up $S \geq 1.84$ compared to the base case). As opposed to the **Merger** benchmark, the tree-based applications require unrolling of one or two loop iterations ($PF$) until disjointness of sub-structures can be determined (Sect. 4.2.1) which explains the resource overhead compared to the base case (especially noticeable in DSP slice consumption). All other tree-based benchmarks require one loop iteration to be peeled off before the parallelisation is successful.

For the benchmarks **Merger** and **Filter**, we include an additional case study by adding two reference designs for comparison shown in Table 4.2: hand-optimised HLS designs using Vivado HLS and hand-written RTL designs in VHDL. For **Filter**, these are the optimised HLS and RTL designs in Chap. 2. The manual HLS design of **Merger** achieves a slightly lower cycle count, but only a slightly lower clock rate. This results in a faster design for the automatically parallelised case. The manual RTL design has the lowest cycle count and execution time. Comparing resources, clock frequency and cycle count for **Filter**, we observe further improvements obtained from manual source code refactoring (∼2.5× faster): in the hand-optimised HLS design, we manually flattened loop nests in order to enable efficient pipelining of the tree traversal loop (Sect. 2.4.2), an optimisation beyond the scope of our automated transformation. This loop contains two sub-loops with variable bounds and code at each loop-level. It is not a perfectly or semi-perfectly nested loop, which prevents the application of the Vivado HLS loop flattening directive. Without loop flattening, only the inner loops can be pipelined, which results in less speed-up compared to the manually flattened loop. The manual HLS design remains more than 3× slower than the RTL implementation because the tree traversal must be distributed over a producer and a (flattened) consumer loop, while it is implemented in a single pipeline in the RTL design (Sect. 2.3.2). Furthermore, the use of bit width customisations of data items and pointers in the manual designs, which reduces the memory consumption, is beyond the scope of this work.

We perform an evaluation of the tool execution time on a machine with an Intel i7-3770 processor (3.40 GHz) and 16 GB memory. The heap analyser consumes the majority of the overall execution time, which varies significantly across our benchmarks. Table 4.3 shows the analysis time broken down into cut-point insertion and fix-point calculation. The latter dominates the run-time and is very sensitive to the number of disjunctive clauses in our state formulae that arise from branching instructions in the program. Columns 2 and 3 show the number of fix-point iterations

**Table 4.2** Comparison with hand-written HLS/RTL designs

*P*: parallelisation degree; *PF*: peeling factor (number of initially unrolled iterations); *S*: speed-up over baseline

| | P | LUT | FF | DSP | BRAM | Clock rate | Latency | *S* |
|---|---|---|---|---|---|---|---|---|
| **Merger** (4 × 2048 random input key-value pairs) | | | | | | | | |
| Baseline (reference) | 1 | 1644 | 1547 | 0 | 96 | 143 MHz | 88.8 ms | 1.00 |
| Autom. parallelisation (**this work**) | 4 | 2012 | 1901 | 0 | 82.5 | 198 MHz | 16.7 ms | **5.31** |
| Hand-written HLS (Chap. 2) | 4 | 1392 | 1253 | 0 | 60 | 173 MHz | 19.3 ms | 4.59 |
| Hand-written RTL (Chap. 2) | 4 | 1462 | 1833 | 0 | 52 | 200 MHz | 11.0 ms | 8.08 |
| **Filter** (16384 3-dim. data points, 32767 tree nodes, $K = 128$ clusters) | | | | | | | | |
| Baseline (reference) | 1 | 8387 | 4981 | 18 | 609.5 | 181 MHz | 5390.6 us | 1.00 |
| Autom. parallelisation (**this work**) | 2 | 14197 | 12145 | 72 | 614.5 | 200 MHz | 2860.3 us | **1.88** |
| Hand-written HLS (Chap. 2) | 2 | 15046 | 13612 | 36 | 507 | 182 MHz | 902.0 us | 5.98 |
| Hand-written RTL (Chap. 2) | 2 | 10418 | 19008 | 40 | 448 | 200 MHz | 270.5 us | 19.93 |

**Table 4.3** Tool execution time

| | Fix-point iterations | Average number of disjunctive clauses per iteration | Cut-point insertion (s) | Fix-point calculation (s) | Total (analysis) | HLS + RTL Implementation (s) |
|---|---|---|---|---|---|---|
| Merger | 2 | 1.0 | 1.6 | 1544.5 | 1546.1 | 300.0 |
| Tree Deletion | 3 | 3.3 | 2.2 | 4.0 | 6.2 | 543.1 |
| Filter | 7 | 10.3 | 23.5 | 322.1 | 345.6 | 1006.3 |
| Reflect Tree | 3 | 3.3 | 2.3 | 4.2 | 6.5 | 353.6 |

required for a disjointness proof and the average number of disjunctive clauses per iteration, respectively. The analysis time for **Tree Deletion** and **Reflect Tree** is short because a fix-point is reached quickly. **Merger** does not need more fix-point iterations, but a single symbolic execution of the loop body is slow because each symbolic execution of an outer loop iteration must converge to a fix-point for the inner loop. **Filter** requires 7 fix-point iterations with an average number 10.3 disjunctive clauses per iteration. Figure 4.6 (left) shows the total (net) number of clauses per iteration and the number of removed clauses due to merging for **Filter**. The right figure shows the execution time per fix-point iteration.

## 4.5 Performance and Robustness of the Heap Analysis

The heap analysis is the core element of our framework. We discuss its performance and its relation to previous work, and identify weaknesses which motivate future research. An advantage of our technique is that it can, beyond straight-line code and

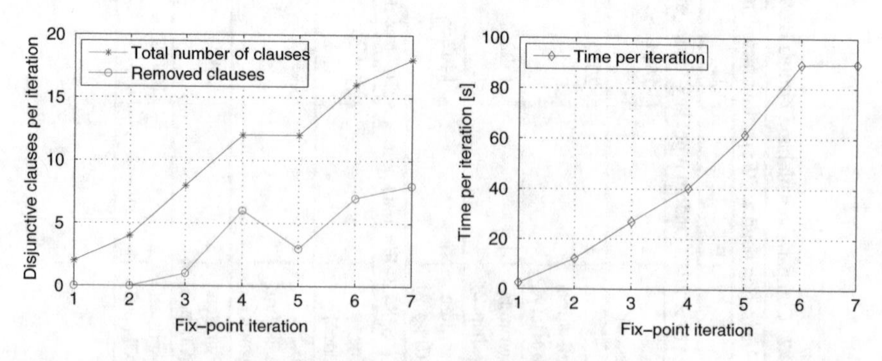


**Fig. 4.6** Analysis complexity for **Filter**. *Left* number of disjunctive clauses (total and removed). *Right* tool execution time per fix-point iteration

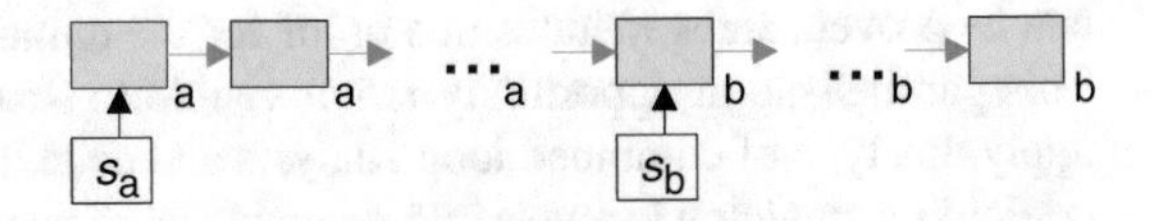


**Fig. 4.7** Punctured linked list

deterministic static control parts such as unrollable `for`-loops, handle `while`-loops enclosing data-dependent conditionals, and with data-dependent loop condition and unknown iterations count. This feature distinguishes our analysis from the polyhedral model which is based on a full enumeration of the iteration space at compile time. On the other hand, this feature requires us to describe data structures of unknown size to ensure convergence of the fix-point calculation. We achieve this with recursive predicates in separation logic discussed in Sects. 3.5.1 and 4.2.2, which allow us to describe pointer-linked data structures with a compact formula. Describing binary trees in classical logic is much more long-winded. For example, if one was to describe the tree predicate from Definition 3.3

$$\mathit{tree}(E) \iff (\, E = \texttt{nil} \wedge \mathit{emp}) \vee (E \mapsto [\texttt{l} : x_1', \texttt{r} : y_1'] * \mathit{tree}(x_1') * \mathit{tree}(y_1')\,)$$

in classical logic, one must explicitly encode the sharing patterns by imposing conditions on which nodes are reachable from each node:

$$\begin{aligned}\mathit{tree}'(E) \iff\ & (\, E = \texttt{nil}) \vee ([E] = (x_1', y_1') \wedge \mathit{tree}'(x_1) \wedge \mathit{tree}'(y_1') \wedge \\ & E \notin \mathit{reachable}(x_1') \cup \mathit{reachable}(y_1') \wedge \\ & \mathit{reachable}(x_1') \cap \mathit{reachable}(y_1') = \emptyset\,)\end{aligned}$$

where

$$\mathit{reachable}(E) \;=\; \begin{cases} \{\} & \text{if } E = \texttt{nil} \\ \{E\} \cup \mathit{reachable}(x_1') \cup \mathit{reachable}(y_1') & \text{if } [E] = (x_1', y_1'). \end{cases}$$

In contrast to a *reachability analysis* [16], the separation logic-based heap footprint analysis can also partition cyclic data structures, such as non-`nil`-terminated list segments or doubly linked lists because it is based on symbolic execution which mimics the actual program execution and heap accesses. For example, for a program accessing a linked list punctured by two cut-points $s_a$ and $s_b$ as in Fig. 4.7, our analysis determines the disjointness of the list segments labelled $a$ and $b$, which cannot be determined with a reachability analysis as all nodes reachable from $s_b$ are also reachable from $s_a$. The strength of Raza's labelled symbolic execution [3], which our analysis builds on, is thus the detection of the *actually* accessed heap portions, not the portions that may *possibly* be accessed.

While Raza's technique is excellent for proving the independence (with respect to heap-carried data dependencies) between program statements, our analysis is able to reveal more parallelisation and partitioning opportunities. This is because it examines the loop iteration space, peels off a finite number of loop iterations until disjointness can be proven, and generates this proof for the remaining subset of loop iterations. This parallelisation opportunity is not visible to Raza's technique which does not apply this type of combined loop analysis and code transformation.

Folding singleton heaplets into recursive predicates is essential for the successful termination of the loop analysis. For example, our analysis automatically folds

$$s \mapsto [\mathtt{n} : s_1'] * s_1' \mapsto [\mathtt{n} : s_0'] * s_0' \mapsto [\mathtt{n} : \mathtt{nil}]$$

into $ls(s, \mathtt{nil})$. The recursive predicates are defined in logic rules used by the built-in theorem prover which automatically searches for applicable rules. We define a set of predicates for common data structures such as trees, lists, lists with additional pointers to singleton heaplets and sub-trees. These allow us to cover a large range of pointer-based programs. However, we may find applications using more exotic structures for which no folding rule in our current set applies. This limitation can be removed by integrating algorithms for automatic inference of recursive predicates, such as [17], in our tool. The decision under what conditions the folding is triggered builds on a heuristic [1] which 'gobbles up' heaplets by recursive predicates if their pointers are primed variables which do not appear in any other part of the formula except of the predicates involved in the folding. The heuristic works well in practice and we are not aware of a theorem prover implementing a more robust technique. However, in general, we cannot rule out situations where the folding fails due to the incompleteness of the heuristic. A code example where this is the case in given in [1]. In cases where the fix-point calculation does not converge after a pre-defined number of iterations, our tool reports a failed proof and continues the implementation with automatic parallelisation and memory partitioning.

Missing a parallelisation opportunity due to incompleteness also applies for Algorithm 6 itself which uses a heuristic to distinguish private from shared heap regions. Our analysis may thus indicate sharing of a heaplet which in reality is private to a particular code section. In this case, the current analysis aborts. However, in the next chapter, we include the possibility of sharing in our analysis such that this does not result an abortion, but causes the tool to simply instantiate an unnecessary coherency mechanism in this case.

The scalability of the analysis is determined by the fix-point calculation which performs repeated symbolic executions of the loop body until convergence. Non-deterministic branching (e.g. data dependent conditionals) in the loop body results in several disjunctive clauses describing the loop state as all control flow paths must be analysed. In the worst case the number of these clauses can grow exponentially with the number of fix-point iterations. However, we do not see an exponential growth in our case studies as shown in Fig. 4.6 because our analysis merges equivalent clauses at the end of each iteration.

## 4.6 Summary

This chapter presents a tool flow that automatically parallelises loops in heap-manipulating C/C++ programs and distributes heap-allocated, pointer-linked data structures over separate banks of on-chip block memory in order to leverage the memory-level parallelism in FPGAs. The core of our tool flow is the heap analyser for proving communication-free parallelism in loops. We develop and implement an algorithm for the disjointness/dependence analysis which draws on several existing techniques developed in the separation logic framework: symbolic execution, heap footprint analysis and loop invariant synthesis. The outcome of the analysis is information about the legality of parallelisation and an assignment of heaplets to on-chip memory partitions. The analysis is accompanied by automated code transformations which ensure the synthesisability of the pointer-manipulating program by standard HLS tools, and implement the parallelisation and memory partitioning. Our implementation takes LLVM IR as input, which is generated from a C/C++ program using readily available third-party tools, and produces modified LLVM IR. This output can be used by off-the-shelf LLVM-based HLS tools to generate a hardware description. We demonstrate the successful parallelisation and memory partitioning by our tool flow using four real-life applications and using Xilinx Vivado HLS as an exemplary back-end tool. The HLS implementations parallelised by our tool achieve the expected acceleration by a factor of $1.8\times$ to $5.3\times$ in terms of execution time compared to the non-parallelised implementations. The work discussed above was first published in [12, 18].

The CAD flow described in this chapter performs code optimisations that target the partitioning of on-chip memory resources. However, applications with large memory footprints quickly exceed the on-chip memory capacity and therefore require access to external memory. Off-chip memory access can substantially slow down an FPGA accelerator due to bandwidth limitations. Buffering frequently reused data on chip is a common approach to address this problem and the optimisation of the cache architecture introduces yet another complex design space. The next chapter extends the automatic parallelisation and memory space partitioning technique above to an HLS design aid that generates parallel application-specific multi-cache architectures, hence enabling efficient hardware implementations from memory-intensive pointer-based C/C++ programs.

Apart from the focus on on-chip memories, the above analysis successfully optimises an implementation only if the accessed address space can be fully partitioned into disjoint regions. This excludes applications that inherently share data between parallel functional units. The next chapter also addresses this limitation in that the extended analysis decides whether the parallelisation is legal in the presence of shared resources and guides source code transformations to automatically insert the required synchronisation primitives.

## References

1. S. Magill, A. Nanevski, E. Clarke, P. Lee, Inferring invariants in separation logic for imperative list-processing programs, in *Proceedings of the Workshop on Semantics, Program Analysis, and Computing Environments for Memory Management (SPACE)* (2006), pp. 47–60
2. D. Distefano, P. O'Hearn, H. Yang, A local shape analysis based on separation logic, in *Proceedings of the international Conference on Tools and Algorithms for the Construction and Analysis of Systems* (2006), pp. 287–302
3. M. Raza, C. Calcagno, P. Gardner, Automatic parallelization with separation logic, in *Programming Languages and Systems* (2009), pp. 348–362
4. M. Botinčan, D. Distefano, M. Dodds, R. Grigore, M.J. Parkinson, coreStar: the core of jStar. Boogie 65–77 (2011)
5. Xilinx Vivado HLS, Accessed 12 May 2015. http://www.xilinx.com/products/design-tools/vivado/integration/esl-design.html
6. Altera SDK for OpenCL, Accessed 13 Jan 2016. https://www.altera.com/products/design-software/embedded-software-developers/opencl/overview.html
7. Xilinx SDAccel Development Environment for OpenCL, Accessed 13 Jan 2016. http://www.xilinx.com/products/design-tools/software-zone/sdaccel.html
8. High-Level Synthesis with LegUp, Accessed 20 Oct 2015. http://legup.eecg.utoronto.ca/
9. ROCCC 2.0|Jacquard Computing, Accessed 12 May 2015. http://www.jacquardcomputing.com/roccc/
10. Clang: A C Language Family Frontend for LLVM. http://clang.llvm.org. Accessed 28 Feb 2016
11. The LLVM Compiler Infrastructure. http://llvm.org/. Accessed 28 Feb 2016
12. F.J. Winterstein, S.R. Bayliss, G.A. Constantinides, Separation logic for high-level synthesis. ACM Trans. Reconfigurable Technol. Syst. **9**(2), 10:1–10:23 (2015)
13. ROSE Compiler Infrastructure. http://rosecompiler.org/. Accessed 13 Jul 2014
14. B. Cook, A. Gupta, S. Magill, A. Rybalchenko, J. Simsa, S. Singh, V. Vafeiadis, *Finding Heap-Bounds for Hardware Synthesis, in Formal Methods in Computer-Aided Design* (IEEE, New York, 2009), pp. 205–212
15. T. Kanungo, D. Mount, N. Netanyahu, C. Piatko, R. Silverman, A. Wu, An efficient K-means clustering algorithm: analysis and implementation. IEEE Trans. Pattern Anal. Mach. Intell. **24**(7), 881–892 (2002)
16. L. Hendren, A. Nicolau, Parallelizing programs with recursive data structures. IEEE Trans. Parallel Distrib. Syst. **1**(1), 35–47 (1990)
17. B. Guo, N. Vachharajani, D.I. August, Shape analysis with inductive recursion synthesis. ACM SIGPLAN Notices **42**(6), 256 (2007)
18. F. Winterstein, S. Bayliss, G.A. Constantinides, Separation logic-assisted code transformations for efficient high-level synthesis, in *Proceedings of the IEEE International Symposium on Field-Programmable Custom Computing Machines (FCCM)* (2014), pp. 1–8

# Chapter 5
# Custom Multi-cache Architectures

The extraction of parallelism is crucial for achieving good quality of results. Computational parallelism also requires that the memory system is not a sequential bottleneck to performance. The distributed memory architecture in FPGAs can provide enormous memory bandwidth if the program data can be partitioned and distributed over multiple on-chip memory banks. Parallel on-chip memory capacity remains a scarce resource and many FPGA applications that process large data sets require access to a large off-chip memory. The bandwidth limitations of external memory can significantly slow down an FPGA accelerator and potentially eliminate the gain of parallelisation. An application-specific optimisation of the on-chip/off-chip memory architecture is thus crucial for mapping a program to an efficient FPGA implementation.

Caching frequently reused data is a common approach to reduce the number of expensive accesses to an external memory. FPGAs allow the implementor to tailor such a memory interface according to the requirements of the application. An application-specific optimisation of this architecture introduces yet another complex design space and remains a complex task for a developer. Furthermore, automatic cache design in an HLS context requires the extraction of application-specific properties from program descriptions and remains foreign to most HLS flows. The work presented in this chapter seeks to bridge this gap. We present an HLS design aid that inserts multiple on-chip caches into the interface to an off-chip memory, which results in an application-specific high-performance memory hierarchy. Our technique leverages recent memory abstractions [1, 2], which build an on-chip/off-chip memory hierarchy underneath a uniform interface and which we refer to as *scratchpads* (SPs) in this chapter. Each single SP contains an optional on-chip cache and automatically ensures coherency between the cache contents and data in off-chip memory for an arbitrary memory access pattern [1]. SPs also provide an optional mechanism to maintain coherency between the on-chip caches in multiple, parallel SPs [2]. In this work, we leverage our program analysis to determine whether or not and for which caches an inter-cache coherency mechanism is required in the generated multi-cache



F. Winterstein, *Separation Logic for High-level Synthesis*, Springer Thesis,
DOI 10.1007/978-3-319-53222-6_5

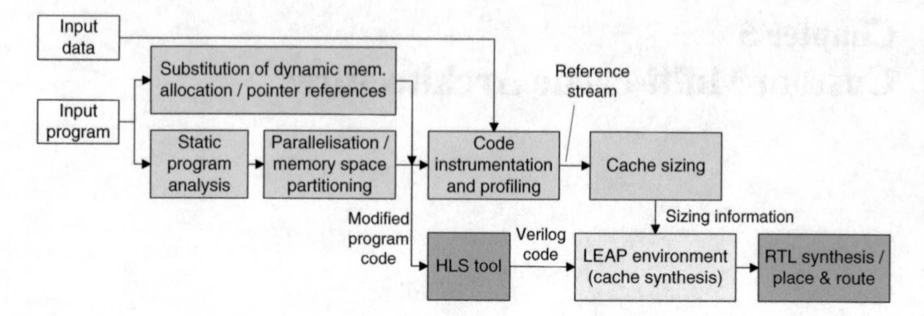


**Fig. 5.1** Summary of the extended tool flow presented in this chapter

architecture. In the following discussion, we refer to caches which require such a coherency network as coherent caches.

This work builds on the static program analysis presented in the previous chapter that extracts memory access information. The applicability of the baseline technique is limited to cases where the on-chip memory capacity is sufficient and the accessed memory space can be split into independent, private partitions. Here, we extend it to shared resources and apply it to the synthesis of efficient interfaces to an off-chip memory. To the best of our knowledge, this is the first application of a separation logic-based analysis to an automated optimisation of the on-chip/off-chip memory hierarchy for FPGA accelerators. Figure 5.1 shows the high-level overview of the extended tool flow described in this chapter. The contributions made in this chapter are:

- In addition to the identification of disjoint heap regions, we extend the baseline analysis in Chap. 4 by an identification of heaplets that would be shared by the parallel loop kernels after parallelisation by the source-to-source translator. Our analysis inserts additional synchronisation primitives for program parts that access shared resources (Sect. 5.2).
- Even if coherency is ensured, updates to the shared resource may happen in a different order after parallelisation compared to the sequential program. This chapter presents a *commutativity analysis* for the shared heap update in order to prove that the parallelisation is semantics-preserving (Sect. 5.2.2).
- The framework targets FPGA accelerators with access to an off-chip memory. The disjointness and sharing information provided by our analyses is used to break the heap (residing in off-chip memory by default) into heaplets, to generate an application-specific parallel multi-cache architecture containing on-chip caches and (if needed) coherency mechanisms; we synthesise parallel private scratchpads for disjoint heap regions and (inherently more expensive) coherent parallel scratchpads for shared regions (Sect. 5.3).
- We extend the cache compilation framework by a dynamic (input data dependent) program analysis to implement an automated size scaling of private caches using spare on-chip memory resources. We include a cache hit rate estimator based on the memory reference trace of the program under test and find the best size distribution

across multiple caches for a user-provided memory access pattern of a particular application (Sect. 5.4).

- We demonstrate the effectiveness of our technique using three applications as test cases which dynamically allocate memory and traverse and update heap-allocated data structures. We use Xilinx Vivado HLS as an exemplary back-end HLS tool in our case studies. We use the open-source LEAP infrastructure [3] and implement our test cases on a Virtex 7 FPGA connected to a DDR3 memory (Sect. 5.5).

## 5.1 Motivating Example

This section reviews a motivating example in the context of the previous chapter and explains how the extensions of the baseline analysis are applied to generate a multi-cache architecture for both private and shared heap regions. Listing 5.1 shows a modified version of the tree-based $K$-means clustering implementation in the previous chapter. The `while`-loop in `filter2` accesses four heap-allocated data structures: the binary tree (type `TR`), the sets of candidate centres (type `CS`), the stack (type `ST`) and the centroid information (type `CI`). Tree nodes, centre sets and stack record are the same data structures as in Chap. 4 (we abbreviate their type identifiers for ease of illustration here). The tree has been built up from the data set to be clustered. The centre sets are intermediate solutions propagated through the call graph. The stack data structure stores the pointers to left and right sub-trees and to the centre sets. The auxiliary functions `push` (Lines 5, 16 and 17) and `pop` (Line 8) are equivalent to the previous example. The difference to Chap. 4 is the centroid information. If the data-dependent conditional (Line 15) evaluates to false (dead end of the tree traversal) the centroid data structure is updated (Lines 22 and 23) which contains the information from which the final clustering result is calculated. As we shall see below, adding this code fragment results in a shared resource after parallelisation of the application.

```
1  //main kernel function
2  void filter2(TR *root, CS cinit, CI *z) {
3    CS* c0 = new CS;
4    *c0 = cinit;
5    ST *s = push(root, c0, true, NULL);
6    while (s != NULL) {
7      TR *u; CS *c; bool d;
8      s = pop(&u, &c, &d, s);
9      CS cs = *c;
10     if (d) {
11       delete c;
12     }
13     CS *cnew = new CS;
14     *cnew = subfunction1(cs);
```

```
15      if (u->left!=NULL) && (u->right!=NULL
          ) && (subfunction2(cs))) {
16        s = push(u->left, cnew, true, s);
17        s = push(u->right, cnew, false, s);
18      } else {
19        delete cnew;
20        // update centroid information
21        CI w = u->wgtCent;
22        CI wprev = z->wgtCent;
23        z->wgtCent = wprev+ w;
24      }
25      delete u;
26    }
27  }
28
29  //auxiliary function push (create new
      entry)
30  inline ST* push(TR *u, CS *c, bool d, ST *
      s){
31    ST *t = new ST;
32    t->u=u; t->c=c; t->d=d; t->n=s;
33    return t;
34  }
35
36  //auxiliary function pop (delete list head
      )
37  inline ST* pop(TR **u, CS **c, bool *d, ST
      *s){
38    *u=s->u; *c=s->c; *d=s->d; ST *t=s->n;
39    delete s;
40    return t;
41  }
```

**Listing 5.1** C-like pseudo code of the $K$-means clustering kernel.

All data structures accessed by this program are created at run-time using dynamic memory allocation. Allocating memory at run-time results in efficient memory usage if the average-case amount of required memory is much smaller than the worst-case amount. An efficient memory architecture for this program provides fast access to this small amount of memory space and, at the same time, supports worst-case allocation by providing a large memory as a backup. Hence, our approach is to place, by default, all heap-allocated data in a large off-chip memory connected to the FPGA accelerator and to insert scratchpads including on-chip caches which mirror parts of the off-chip data and provide fast data access. We describe the extensions of our baseline analysis below.

### 5.1.1 Memory Partitioning and Parallelisation

Figure 5.2 shows an example of the data structures allocated in the heap after executing two `while`-loop iterations. The data structures are grouped according to their types. The loop is split into parallel sub-loops as shown in Listing 5.2 (two in this example). If we ignore the centroid data structure (type `CI`) in the heap layout in Fig. 5.2 for a moment, the baseline method in Chap. 4 can prove that the pointers dereferenced in any iteration of a sub-loop never refer to the data structures used by the other loop. Hence, we call these loop kernels 'communication free' with respect to each other, which satisfies the independence condition that two parts of a program can operate in parallel if they access different data. The analysis partitions the remaining tree data structure (dark grey nodes, type `TR`) into two sub-trees labelled with $\{a\}$ and $\{b\}$. It splits the linked list (type `ST`) into the uppermost node and the nodes below, and the pool of centre sets (type `CS`) is partitioned accordingly. The generation of the multi-cache architecture in this chapter uses the heap partitioning information from the baseline analysis. Each of the parallel sub-loops obtains its own interfaces to off-chip memory and the fact that the memory regions can be proven to be non-overlapping allows our setup to instantiate private SPs for each partition without the need to ensure coherency between them, greatly reducing hardware implementation cost: private caches are faster and cheaper (in terms of FPGA resources) than coherent memory interfaces as ensuring consistency between parallel units with a coherent cache protocol and synchronisation primitives is not required.

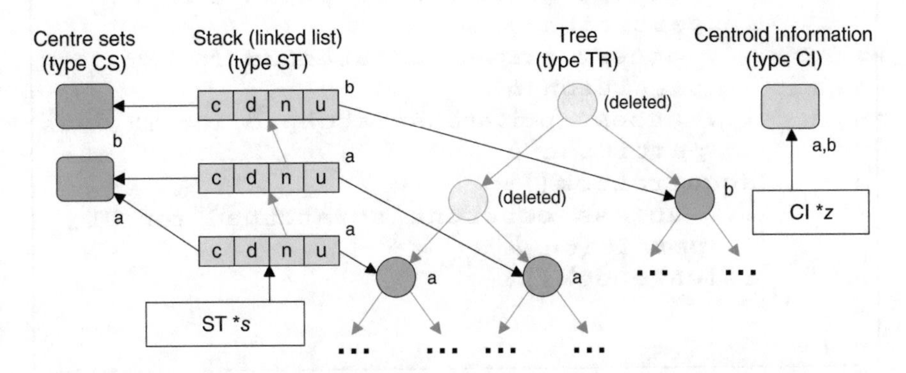


**Fig. 5.2** Snapshot of the pointer-linked dynamic data structures accessed by the loop in Listing 5.1

```
1  //main kernel function
2  void filter2(TR *root, CS cinit, CI *z) {
3
4     ...preamble (pointers access partitions
        a and b)
5
6     while (s != NULL && s != s_b) {
7        //parallel loop kernel a
8        ... access private scratchpad for CS,
           partition a
9        ... access private scratchpad for ST,
           partition a
10       ... access private scratchpad for TR,
           partition a
11       acquireLock();
12       ... access coherent scratchpad for CI,
           partition a
13       releaseLock();
14    }
15
16    s = s_b;
17    while (s != NULL) {
18       //parallel loop kernel b
19       ... access private scratchpad for CS,
           partition b
20       ... access private scratchpad for ST,
           partition b
21       ... access private scratchpad for TR,
           partition b
22       acquireLock();
23       ... access coherent scratchpad for CI,
           partition b
24       releaseLock();
25    }
26 }
```

**Listing 5.2** Transformed program from Listing 5.1.

### 5.1.2 Parallel Access to Shared Resources

Our baseline analysis in the previous chapter cannot handle situations including the shared centroid information in Fig. 5.2. Our extended analysis marks it as a shared resource, indicated by the label $\{a, b\}$, as both sub-loops would update it after parallelisation. After the detection of a shared heap region, our framework instantiates a coherent memory interface [2] to this region in each of the sub-loops. The coherent interface consists of two parts: SPs with caches and a coherence mechanism that ensures data coherency between them, and locks which enable atomic updates of the shared resource in the presence of multiple accessors. The detection of a shared resource triggers a second analysis as sharing invalidates the independence assumption that parallel units access different data. Assuming that coherency is ensured between parallel units, it remains to prove that the modified order in which the shared resource is updated after parallelisation does not alter the program semantics. The centroid information is updated in Line 23 of Listing 5.1:

$$z \rightarrow \texttt{wgtCent} = w_{prev} + w;$$

where $w$ is the contribution of the tree nodes. In the original, sequential program, $z$ receives the contributions of all nodes in the right sub-tree (labelled with $a$) before it receives the first contribution from the left sub-tree (labelled with $b$ in Fig. 5.2). However, in the parallelised version $z$ may be updated with data from left and right sub-tree in an arbitrarily interleaved fashion. Even if atomicity of the update is ensured, we must also ensure that this new update order is legal. In this example, the parallelisation is legal because of the commutativity and associativity of the addition.[1] In general, we address this question with a *commutativity analysis* of the update function.

Listing 5.2 shows the final result of a source code transformation based on the result of all analyses above. The transformed source code, when run through a back-end HLS tool and RTL implementation, results in a custom configuration of multiple private/coherent scratchpads with a custom degree of parallelism. The on-chip memory blocks in the FPGA are aggregated accordingly in order to construct the application-specific parallel caching scheme.

### 5.1.3 Custom Cache Sizing

The above identification of disjoint and shared heap regions and the legality of parallelisation are based on an extension of the static program analysis in Chap. 4. This analysis provides information about the type of the inserted caches, but no information about their size. Hence, all caches inserted by the tool flow above have the same size by default. However, we synthesise a cache for each data structure partition, so the access patterns to these memory regions may be very different: For example, the stack data structure in Listing 5.1 is usually small compared to the tree structure and has high access locality at the head of the stack. In this case, using the available on-chip memory to build a small cache for the stack and a large cache for

[1] We focus on integer or fixed-point systems and ignore non-associativity caused by floating-point representations.

the tree is more beneficial than both caches having the same size. In addition to the static analysis above, we extend our tool flow by a profiling-based program analysis in Sect. 5.4 that aims to construct custom-sized caches to maximise the aggregate hit rate of the multi-cache system. In contract to the static counterpart, we refer to this type of analysis as dynamic analysis.

## 5.2 Extended Static Program Analysis

This section describes the extension of the baseline program analyses enabling source code transformations that turn a sequential heap-manipulating program into a parallelised HLS implementation with an application-specific off-chip memory interface. The baseline analysis for identifying private heap regions and memory partitioning is the starting point for all subsequent analyses related to parallelisation, shared resources and commutativity of shared resource updates. Loop parallelisation and its follow-up analyses are only triggered if (at least parts of) the heap-allocated data structures accessed by the loop can be split into $P$ partitions, where $P$ is the desired degree of parallelism.

The inner `repeat-until`-loop in Algorithm 7 corresponds to the baseline analysis in Chap. 4. It starts with a symbolic execution of the loop preamble and a finite number of the first loop iterations (function SYMBEXELOOPBODY). In each step, it explores the separation logic formula describing the pre-state of the loop $(\Pi \wedge \Sigma_{\{FT\}})$, i.e. the program state before executing the loop body. The algorithm inserts cut-points into the loop pre-state formulae (function CUTPINSERT) while it peels off loop iterations so as to find a valid cut-point assignment. A valid cut-point assignment is found if the built-in proof engine, performing the fix-point calculation for the loop-under-analysis (function FIXPCALC) proves that the initial partitioning of the heap-allocated data structures is maintained for all subsequent loop iterations.

The proof of loop-invariant resource separation is generated by assigning a state to each inserted cut-point (function ASSIGNCPSTATES). The fix-point calculation assigns footprint labels to the accessed heaplets according to the current cut-point state, which changes once a heaplet referenced by a different cut-point is accessed during the symbolic execution. Complete partitioning of the heap accessed by the loop-under-test is proven by the absence of non-singleton label sets attached to the heaplets in the state formulae. If we ignore the centroid information in the motivating example, starting from the pre-state in Fig. 5.2 and with a second cut-point $s_b$ (in addition to $s$) referencing the uppermost stack record in Fig. 5.2, the proof of complete separability is generated by our baseline analysis, a prerequisite for parallelisation.

**Algorithm 7** Detecting Private and Shared Resources.

```
 1: Input:
 2: loop body specification (code)
 3: initial state formula (Π ∧ Σ_{FT})^initial (from symbolic execution of loop preamble)
 4: parallelisation hypothesis P
 5: Output:
 6: number of initial unrollings required (it)
 7: label mapping: program statement identifiers to heap partitions (m)
 8: set of statement sets accessing shared heaplets (StmtS) from which private/shared predicates
    of memory interfaces can be derived
 9: Variables:
10: it                        ▷ Iteration counter (number of iterations to be unrolled)
11: C                                                          ▷ set of cut-points
12: C_shared                          ▷ set of cut-points referencing shared heaplets
13: S_cutpoints                                            ▷ set of cut-point states
14: Π ∧ Σ_{FT}     ▷ state formula in separation logic (attached footprint label set FT)
15: Π ∧ Σ_{CS}     ▷ state formula in separation logic (attached cut-point state set CS)
16: m                                                ▷ label mapping m: FT → CS
17: StmtS                       ▷ set of statement sets accessing shared heaplets

18: function HEAP- ANALYSIS
19:   C_shared ← ∅
20:   repeat
21:     it ← 0
22:     C ← ∅
23:     Π ∧ Σ_{FT} ← (Π ∧ Σ_{FT})^initial
24:     StmtsS ← ∅
25:     success ← false
26:     repeat
27:       while not checkIfValidCutpInsertion(Π ∧ Σ_{FT}, C) do
28:         Π ∧ Σ_{FT} ← SymbExec(Π ∧ Σ_{FT}, it)              ▷ peel off it iterations
29:         Π ∧ Σ_{FT}, C ← CutpInsert(Π ∧ Σ_{FT}, P, C_shared)  ▷ insert P cut-points
30:         it ← it + 1
31:       end while
32:       S_cutpoints ← AssignCPStates(C)                   ▷ assign states to cut-points
33:       Π ∧ Σ_{CS}, success, Stmts_shared ← FixpCalc(Π ∧ Σ, C, S_cutpoints, C_shared)
34:       m ← GetLabelMapping(Π ∧ Σ_{CS}, Π ∧ Σ_{FT})                     ▷ label mapping
35:       StmtsS ← StmtsS ∪ {Stmts_shared}                       ▷ collect shared accesses
36:     until success or it ≥ L_max
37:     if it = L_max then
38:       C_shared ← ExtrctCutp( argmin_{Stmts∈StmtsS} |Stmts|)
39:     end if
40:   until success
41:   return StmtsS, it, m
42: end function
```

### *5.2.1 Detecting Private and Shared Resources*

The baseline analysis is aborted reporting a failed proof after a fixed parameter of $L_{\max}$ unrollings if the program state cannot be completely partitioned. Here, we relax the constraint that the inherent parallelism of the application needs to be communication-free. Algorithm 7 shows the extended analysis to identify disjoint and shared resources. If we include the centroid information in our motivating example and run the disjointness analysis, the proof engine always finds a non-singleton label set attached to it and never reports a valid proof. Our goal is to mark this heaplet as a shared resource. The shared resource analysis requires two extensions of the baseline analysis: (1) identifying shared heaplets and (2), once marked as shared, re-running the cut-point insertion and proof-engine invocations while excluding them from the search for separable heap regions.

In the first phase, we turn a failed proof of complete separability into the detection of shared resources. We run the cut-point insertion and fix-point calculation with the objective of splitting the heap into $P$ partitions, as shown in the inner `repeat-until`-loop. After peeling off the first loop iteration of the motivating example, the function FIXPCALC terminates unsuccessfully because it finds non-singleton label sets attached to a centre set and the centroid information. After unrolling two iterations, the sharing of a centre set disappears and the centroid information remains as the only shared resource. We use a heuristic approach to filter shared resources by declaring all heaplets having a non-singleton label set after $L_{\max}$ unrollings as shared. The fix-point calculation is modified in that whenever it detects sharing on a heaplet, it collects the set of program statements that accessed the shared heaplet (each statement in the control flow graph has a unique identifier). During the course of the alternating iteration unrolling, cut-point insertion and fix-point calculation, the analysis builds a set of statement sets accessing shared heaplets ($StmtsS$).

After termination of the inner `repeat-until`-loop, the analysis is reset. From $StmtsS$, we pick the set $Stmts$ containing the fewest statements accessing shared resources, from which the function EXTRCTCUTP extracts all cut-points mentioned in at least one of these program statements ($C_{shared}$). The second phase begins by relaunching the analysis. We pass the set $C_{shared}$ to the modified function CUTPINSERT which excludes these cut-points during the search for cut-points in the loop pre-state. Similarly during the fix-point calculation we prevent the analysis from adding a partition label to a heaplet if the current program statement has been marked as excluded. Finally, we obtain a proof of separability for the tree, the stack and the pool of centre sets, and the centroid heaplet is marked as a shared resource. The interface to the shared heap region residing in off-chip memory is then supported by a coherency protocol. The corresponding program statements accessing the shared resource are in Lines 22 and 23. Our analysis extracts these statements and the source code transformation inserts `acquireLock` and `releaseLock` commands before and after the critical statements as shown in Listing 5.2 in order to ensure atomic updates of the shared heap region.

### 5.2.2 Commutativity Analysis

Parallelisation in the presence of shared resources requires a second analysis step after detection of a shared heap region. We must verify that, after parallelisation, the program semantics are not altered as a results of the order in which the updates of the shared resource are made by the parallel version being altered. For example, during the execution of the original (unparallelised) loop in Listing 5.1, the shared centroid information receives all contributions from the right sub-tree before it receives any contribution from the left sub-tree, while it may be updated with data from left and right sub-tree in an arbitrarily interleaved fashion in the parallelised version. Enforcing the original order with barrier synchronisation means re-sequentialising the parallelised implementation and is not a viable solution. Instead we want to determine whether the modified order of state updates is legal. In the following walk-through, for ease of explanation, we define the function $F$ which reads and writes the shared state (Lines 22 and 23 in Listing 5.1):

**Definition 5.1** (*Update function*)

**function** $F(w)$

$w_{prev} = z \rightarrow \texttt{wgtCent};$

$z \rightarrow \texttt{wgtCent} = w_{prev} + w;$

**end function**

A commutativity analysis was proposed by Rinard and Diniz [4] and our approach builds on the same basic idea: we say two operations on the program state are commutable if their execution in sequence results in the same program state regardless of their execution order. In our case, $F$ is commutable if $\forall w_1, w_2,\ F(w_1); F(w_2)$ results in the same program state as $F(w_2); F(w_1)$. From the symbolic execution and detection of the shared resources as above, we extract the pre- and post-conditions on the program state:

$$\begin{aligned}
&\{w = w_0' \wedge z \mapsto [\texttt{wgtCent} : w_1']\} \\
&F \\
&\{w = w_0' \wedge w_2' = w_1' + w_0' \wedge z \mapsto [\texttt{wgtCent} : w_2']\}
\end{aligned} \tag{5.1}$$

The extraction phase brings the pre- and post-specification of $F$ into a canonical form $\Pi \wedge \Sigma$, where $\Pi$ are the pure formulae and $\Sigma$ are the spatial formulae referring to the shared heap resource. For example, the built-in symbolic execution engine ensures that arithmetic operations in the state formulae appear only in the pure part by creating a fresh primed variable $w_2'$. We test whether $F$ is commutable by symbolically executing two sequences of two calls to $F$:

$$w = w'_{0,1};\ F(w);\ w = w'_{0,2};\ F(w);\ w = w'_{0,3}; \tag{5.2}$$

$$w = w'_{0,2};\ F(w);\ w = w'_{0,1};\ F(w);\ w = w'_{0,3}; \tag{5.3}$$

Note the permuted assignment of symbolic values to $w$ in (5.3). In order to show that $F$ is commutable, we must prove that the post-states of the sequences in (5.2) and (5.3) describe the same program state. Their post-state formulae are:

$$w = w'_{0,3} \wedge w'_3 = w'_1 + w'_{0,1} + w'_{0,2} \wedge z \mapsto [\texttt{wgtCent} : w'_3] \tag{5.4}$$

$$w = w'_{0,3} \wedge w'_4 = w'_1 + w'_{0,2} + w'_{0,1} \wedge z \mapsto [\texttt{wgtCent} : w'_4] \tag{5.5}$$

The updated shared resource in (5.4) and (5.5) is described by $z \mapsto [\texttt{wgtCent} : w'_3]$ and $z \mapsto [\texttt{wgtCent} : w'_4]$, respectively. We want to prove that these predicates describe the same state. We first ask a separation logic theorem prover whether they match which recognises their equality in shape and creates a new proof obligation:

$$w'_3 = w'_4 \tag{5.6}$$

Next, we combine the verification condition (5.6) with the remaining pure parts of the formulae and aim to prove:

$$\begin{aligned} &\forall w'_{0,2}, w'_{0,1}. \\ &w = w'_{0,3} \wedge w'_3 = w'_1 + w'_{0,1} + w'_{0,2} \wedge \\ &w = w'_{0,3} \wedge w'_4 = w'_1 + w'_{0,2} + w'_{0,1} \Rightarrow (w'_3 = w'_4) \end{aligned} \tag{5.7}$$

In the actual verification step, we use satisfiability modulo theories (SMT) solving [5] to decide (5.7). However, an SMT solver cannot deal with the universal quantification ($\forall$), so we rephrase (5.7) by negating the verification condition:

$$\begin{aligned} &\exists w'_{0,2}, w'_{0,1}. \\ &w = w'_{0,3} \wedge w'_3 = w'_1 + w'_{0,1} + w'_{0,2} \wedge \\ &w = w'_{0,3} \wedge w'_4 = w'_1 + w'_{0,2} + w'_{0,1} \wedge (w'_3 \neq w'_4) \end{aligned} \tag{5.8}$$

The solver returns one of three possible results: (1) If (5.8) is satisfiable, we can find an assignment to the input variables $w'_{0,2}, w'_{0,1}$ of $F$ that makes the program states after executing both sequences different: $F$ is not commutable. (2) If (5.8) is not satisfiable, there is no such assignment: $F$ is commutable. (3) The solver may not be able to decide the question in which case we conservatively assume that $F$ is not commutable. For the running example and with the theory of linear arithmetic of integers it decides that $F$ is commutable. Commutativity has been shown to be an undecidable problem in general [6]. However, it can still be shown for many cases that arise in practice.

The next section describes our compilation flow that uses the information provided by the above program analyses to generate application-specific multi-cache architectures.

## 5.3 Code Generation

The tool flow implementation of the multi-cache synthesis consists of three main parts: (1) The analysis extension builds on the baseline heap analyser from Chap. 4. It also interfaces the Z3 SMT solver [5]. (2) The modified source-to-source translator builds on the baseline infrastructure which implements the loop parallelisation and pointer access transformations. The code generation now includes directives for instructing Vivado HLS to generate bus interfaces for memory access. (3) We leverage the open-source LEAP (Latency-insensitive Environment for Application Programming) framework [3] to embed the C/C++-based HLS kernels in an environment that provides access to a physical FPGA device and memory.

Like an operating system, LEAP provides a unified layer of abstraction on top of device-specific drivers that interface the underlying FPGA device, on-board memory and the host system into which an FPGA card is plugged. In particular, our setup uses LEAP's *scratchpads* (SPs), a memory interface abstraction for FPGA applications. SPs provide a simple read-request, read-response and write memory interface to the connected application. Internally, LEAP scratchpads instantiate a memory hierarchy: an optional on-chip cache, board-level off-chip memory and finally the main memory of the attached host system as shown in Fig. 5.3. SPs without on-chip caches forward all requests to off-chip memory which results in longer response times. The same applies for cache misses. Evicted items are automatically flushed to the next memory level. The framework provides two types of SPs: (1) *Private scratchpads* [1] are instantiated when memory spaces are known to be disjoint from all regions accessed by other memory interfaces. (2) If several memory interfaces refer to a shared memory region we instantiate *coherent scratchpads* [2]. The latter feature consists of distributed caches backed by a coherence protocol. Multiple coherent SPs appear as independent interfaces to the application, while they are internally connected via a ring network that ensures inter-cache coherency. The shared memory abstraction by coherent scratchpads hides the internals of the coherency mechanism. Coherent SPs are more expensive (in terms of FPGA resources) and slower (in terms of response time) than their private counterparts. Although we build our tool on top of LEAP in this work, we believe that our analysis and code generation framework can be ported to memory abstractions other than LEAP scratchpads, such as CoRAM [7].

The source-to-source transformation replaces heap memory with arrays located in off-chip memory by default (a portion of them then resides on-chip via caches) and each heap access becomes an access to the external memory bus. The translator turns pointer dereferencing into array-based bus accesses and instantiates a memory interface for each data structure type and each of the $P$ heap partitions (private and shared). The extended heap analyser provides information on whether the memory

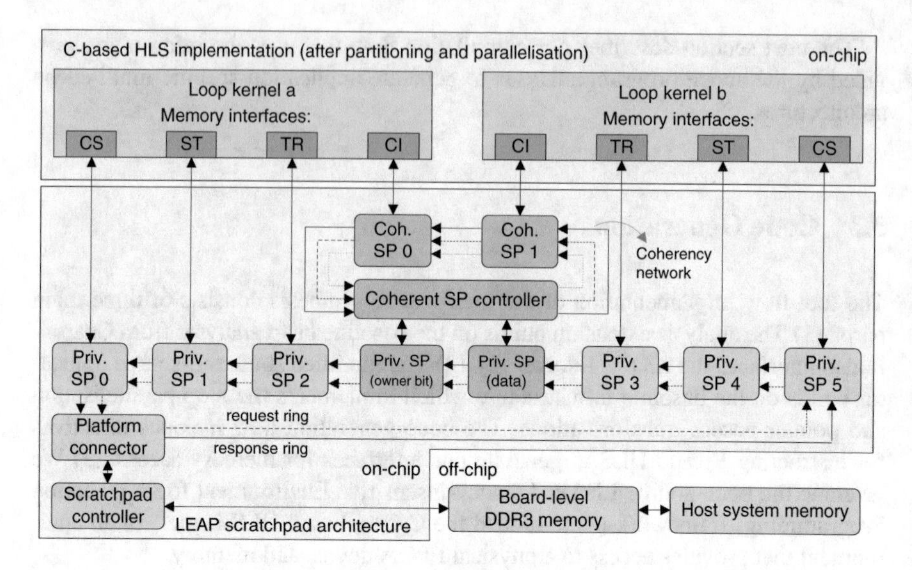


**Fig. 5.3** Parallelised HLS implementation of the filtering algorithm with a hybrid cache architecture

bus points to a private or a shared heap region. We insert a generic Verilog wrapper for each interface which acts as a bridge between Vivado's native bus protocol and the LEAP memory interface. Vivado's scheduler ensures that, when the HLS kernel issues a memory request, it stalls execution until the memory request has been serviced by the SP.

Figure 5.3 shows the integration of our running example after heap partitioning and parallelisation with $P = 2$ into the LEAP framework and memory hierarchy. Each loop kernel (we omit the preamble here) has an interface to the memory system for each type of heap-allocated data-structure: centre sets (CS), stack records (ST), tree nodes (TR) and centroid information (CI). An additional coherency network is instantiated for the CI ports (shared memory). For shared heap regions, the source translator inserts synchronisation signals in order to ensure fine-grain atomic updates to the shared heap cell. Listing 5.3 shows an example. The pass-by-reference argument `access_critical_region0` translates into a Boolean signal in the generated RTL code and triggers lock acquisition and release. The lock service provided by LEAP ensures that no access to heap region 0 is granted before the lock is acquired (only one requestor can own the lock). The memory fence instruction ensures that the memory transaction has been completed before releasing the lock.

The on-chip caches of the private and coherent scratchpads are direct-mapped with write-back policy. The presence of a coherency mechanism is the only variable parameter in our cache architecture implementation above. In particular, we fixed the cache size to 1 kB with 64 bit line size by default. The next section describes an extension of the work above, which replaces the default cache sizing by variable, custom cache sizing.

```
1 requestLock(access_critical_region0);
2 waitForLock(); //stalls until lock has
    been acquired
3 ...issueMemoryRequest //set memory fence
4 releaseLock(access_critical_region0);
```

**Listing 5.3** Lock-synchronised shared memory access.

## 5.4 Custom Cache Sizing

In applications with large memory footprints, such as the application this chapter targets, the bulk of the data necessarily resides off-chip. In these cases, the HLS core often keeps only small data structures on-chip. Consequently, the amount of on-chip BRAM used by the core is often smaller than the amount of the BRAM available. We extend the cache compilation flow in the previous sections by an add-on that automatically uses up the left-over BRAM and enlarges the private on-chip caches. Secondly, the size of each private cache is set individually in order to obtain a size distribution across the parallel caches that is tailored to the memory access pattern of a particular application. For example, the accesses to some data structures of an application may have good locality and increasing the cache size may therefore improve performance. On the other hand, some memory access traces in an application may have very little locality or access small data structures, so scaling up the cache beyond a certain size is of no use. In such a case, an application-specific cache sizing will very likely have superior performance compared to a one-size-fits-all solution.

It is important to note that our technique does not rely on successive synthesis and place-and-route cycles, but instead estimates the cache performance for different sizes with a pre-RTL, dynamic program analysis of the input code to an HLS tool. Our approach relies on a prediction of the performance of each cache from the application's reference stream, and finds a size configuration that maximises the aggregate performance subject to a resource constraint. Although statically determining the cache size requirements and hence the size of the data structures is possible in some corner cases [8], we adopt here a run-time profiling approach for capturing the memory reference trace in order to ensure wide applicability. Especially in heap-manipulating programs, the absolute data structure size is often unknown at compile time. Our dynamic analysis can handle this type of program at the expense of relying on a representative input data set provided by the user.

To give a more concrete motivating example of our technique, we consider a two-cache system consisting of private caches. Our compilation flow above generates such a system, for example, from applications which use a tree data structure and a stack to implement a depth-first tree traversal. The **Reflect Tree** benchmark from Chap. 4 is an example of such an application. Assuming we have only run the transformation of pointer references and cache insertion for **Reflect Tree** without asking for additional parallelisation, the hardware implementation has a private cache for stack records (ST) and tree nodes (TR). The RTL design for the modified source code is generated with an HLS tool, for example Xilinx Vivado HLS, which also provides information of the BRAM resources consumed by the HLS core itself. In this case, the core uses 112 36k-RAM blocks which leaves 918 left-over blocks in a Virtex 7 device (xc7vx485tffg1761-2) to be used by the platform surrounding the HLS core. With a conservative 40%-margin, 550 RAM blocks (2200 kB[2]) can be repurposed as cache memories.

Our technique then estimates the performance of the caches from the memory reference trace, which is obtained from running the HLS input program with a representative input data set provided by the user. The reference stream, together with the knowledge of the cache type (direct-mapped, set-associative, fully associative) allows us to model the aggregate hit rate of the multi-cache system. For $K = 2$ private caches as in this example, there is no interaction between the caches and the aggregate hit rate is given by:

$$\eta = \frac{\sum_{i=0}^{K-1} h_i(B_i)}{\sum_{i=0}^{K-1} t_i}, \tag{5.9}$$

where $h_i$ is the number of hits in cache $i$ of size $B_i$, $t_i$ is the total number of accesses to cache $i$. Figure 5.4 shows the aggregate hit rate for the two direct-mapped caches over different feasible size configurations. The design space spans hit rates from 79 to 97%. The hit rate of Cache 0 (for stack records of type `ST`) reaches its maximum at a size of 32 kB and then plateaus. The reason for the steep improvement with low sizes and early saturation is the high locality of the memory accesses made to the stack-like linked list and the fact that just 32 kB of cache memory is sufficient to keep the entire data structure on-chip. For tree nodes (Cache 1), a 2 MB cache is needed to fit all tree data. Clearly, spending the same amount of memory resources on both caches is sub-optimal.

The advantage of our technique over a one-size-fits-all cache scaling becomes obvious when we take the memory resource constraint of 2200 kB into account. With a fixed size for all caches, on this grid, we could implement caches with a maximal capacity of 1024 kB each, which corresponds to the bar marked with the solid-line blue ellipse in Fig. 5.4. A cache sizing tailored to the access pattern of the application allows us to decide that a size of 32 kB for Cache 0 and 2048 kB for Cache 1 maximises the hit rate while still satisfying the resource constraint. This design point is marked with the dashed blue ellipse in Fig. 5.4. In general, implementations (including those parallelised by our CAD flow) will use more than two parallel caches, and the disparity between fixed-size and application-specific cache sizing will be larger.

Replacing a fixed-size scaling with a specific size distribution relies on the ability to predict the performance of each cache from the application's reference stream, and

[2] We use 32 kbits in a Xilinx 36K-RAM block to store user data.

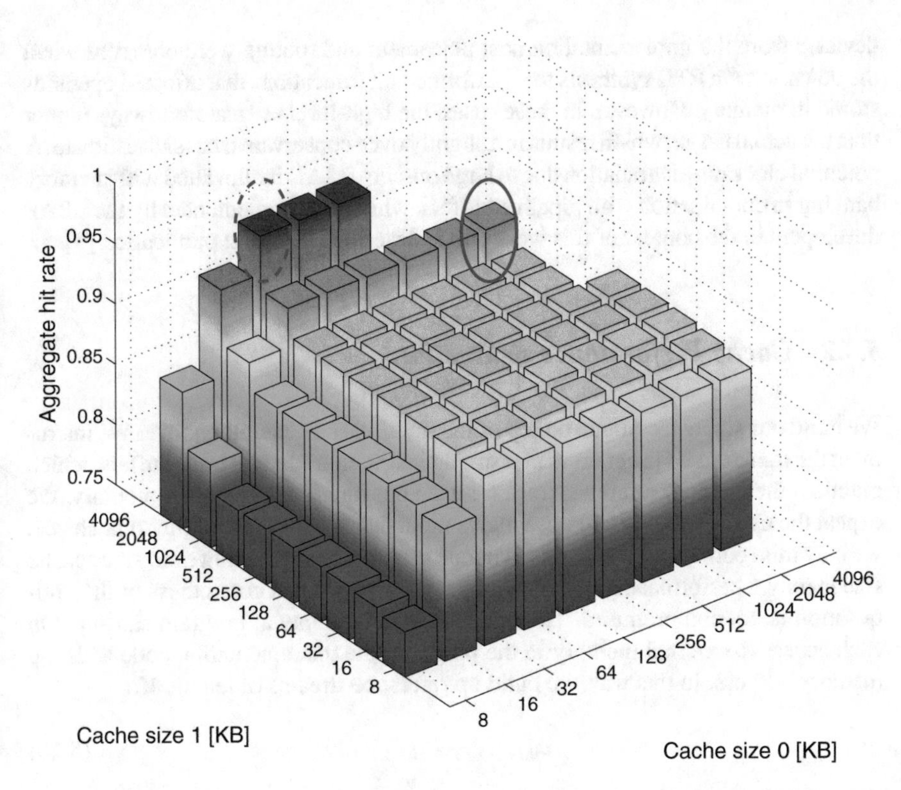


**Fig. 5.4** Aggregate hit rate estimate for a two-cache system with an 2200 kB on-chip memory constraint

to find a cache size configuration maximising the aggregate performance subject to a memory resource constraint. Our cache sizing flow has three components: (1) It first determines unused BRAM resources, which requires an estimation of the memory resources used by the HLS core itself. (2) We predict the hit/miss counts of each cache for different sizes. (3) The amount of spare BRAM and the cache performance estimates are combined into an optimisation problem which finds a variable size configuration in the multi-cache system that maximises the aggregate hit rate.

## 5.4.1 *On-Chip Memory Utilisation Estimation*

We obtain high level estimates of the BRAM consumption from the HLS tool to determine the left-over RAM resources. Here, we use Vivado HLS, which provides estimates of the number of LUTs, FFs, DSP slices and RAM blocks consumed by the HLS core. Compared to LUTs, FFs and DSP slices, the predicted amount of memory is relatively accurate. The only cases where the high-level prediction deviates from the implementation post placement and routing were observed when the down-stream RTL synthesis tool performed bit truncations that affected operands stored in memory. However, in these cases, the high-level estimate is always higher than the actual usage, which results in a slightly over-conservative but safe estimate. A potential clock rate degradation due to large on-chip RAMs is alleviated with memory banking in combination with pipeline buffers, which was implemented by the LEAP developers in the context of this work and is described in a joint publication [9].

### 5.4.2 Cache Performance Estimation

We build our sizing technique on top of the multi-cache generator above. We instrument the transformed program with profiling instructions that fill trace buffers, which maintain the memory reference trace for each bus interface to external memory. We expect the user to provide a representative input data set for the profiling run. Hence, we may miss corner cases with this dynamic program analysis. However, since cache size is only a performance-related parameter, the functional correctness of the optimisation is not compromised. The trace buffers are empty at program start-up. On each access to external memory in the program, the instrumentation code adds the memory address. In this way, we build up reference streams of length $M_i$:

$$( a_{0,i}, \ \ldots, \ a_{M_i-1,i} ), \tag{5.10}$$

where $i$ is the index of the memory interface. The memory is divided up into *blocks*, some of which will have copies in the cache. The block width $L$ is equal to the cache line size. For a data width smaller than $L$ the *block reference streams*

$$( \lfloor \tfrac{a_{0,i}}{L} \rfloor, \ \ldots, \ \lfloor \tfrac{a_{M_i-1,i}}{L} \rfloor ) \tag{5.11}$$

give us the dynamic trace of memory accesses at the granularity of the cache line size. The cache line size is a fixed parameter in our analysis. If the user data width is larger, a cache access is split into multiple sequential chunks in our implementation. We model this by expanding the block reference stream (5.11) accordingly in a post-processing step. The cache size remains the only variable parameter in the hit rate estimation. Other parameters such as associativity and support for disjoint/shared memory accesses are fixed but must be taken into account. Our current cache sizing flow targets private caches only and we discuss the extension to coherent caches in the outlook section of this thesis.

The hit rate of fully associative caches can be precisely determined using the *stack distance* metric [10–13], which counts the number of unique references 'between' accesses to the same address. A cache with $B$ lines then filters out references with stack distance larger than $B$. The stack distance distribution of a reference stream allows us to count cold misses (cache misses due to empty cache at program start-up) and capacity misses (misses due to line eviction because the cache is full) in fully

**Algorithm 8** Hit rate of a private, direct-mapped cache.

```
1: Input:
2: Block reference stream S
3: Number of cache lines B
4: Output:
5: Miss count n_miss
6: Hit count n_hit
7:
8: function ESTIMATE_HITRATE(S)
9:    S_u ← unique(S)                                  ▷ keep unique block references
10:   n_miss, n_hit ← 0
11:   for all r ∈ S_u do
12:      I ← findAll(S = r)                            ▷ get indices of entries equal to r
13:      c ← r mod B                                   ▷ cache line accessed by r
14:      n_miss ← n_miss + 1                           ▷ first access is always a cold miss
15:      for j = 1 ... length(I) − 1 do                ▷ loop over remaining accesses
16:         R' ← S(I(j − 1) + 1 : I(j) − 1)            ▷ intervening refs
17:         C' ← R' mod B                              ▷ intervening cache line refs
18:         if find(C' = c) = ∅ then
19:            n_hit ← n_hit + 1                       ▷ hit
20:         else
21:            n_miss ← n_miss + 1                     ▷ conflict miss
22:         end if
23:      end for
24:   end for
25:   return n_miss, n_hit
26: end function
```

associative caches. In lower-associativity caches, additional conflict misses occur (eviction due to intervening references although the cache is not full) which the stack distance approach can only approximate [12, 13]. The prediction accuracy worsens with decreasing associativity.

Because we target direct-mapped caches and because our goal is an accurate prediction, we devise a precise hit rate determination for direct-mapped caches. For each reference $r$ and the previous reference $r'$ to the same block address, we examine the intervening references made between $r'$ and $r$. A conflict miss occurs if at least one intervening reference accesses the same cache line, which is determined with a modulo operation using the cache size $B$ as divisor. Algorithm 8 shows Matlab-like pseudo code of the hit rate estimator for direct-mapped caches of size $B$. It predicts the number of hits ($n_{\text{hit}}$) and misses ($n_{\text{miss}}$) of the cache dependent on its size, which allows us to compare the performance of cached memory interfaces with different block reference streams) relative to the other caches and select a configuration of cache sizes that maximises the aggregate hit rate. The next section describes how our technique finds such a configuration.

### 5.4.3 *Optimisation Strategy*

Our compiler generates $K$ caches as described above. With Algorithm 8, we can estimate the performance of each independent cache $h_i(B)$, $i = 0 \dots K-1$ once we have obtained the corresponding reference streams. We assign different sizes to the caches in such a way that the aggregate hit rate is maximised. To this end, we assign to each cache a set of $N$ cache sizes $\mathcal{B}_i = \{B_0, B_1, \dots, B_{N-1}\}$ and compute the hit rate relative to the total number of accesses for each size. We cast the search for the best size assignment for each cache into an optimisation problem and define the following variables:

| | |
|---|---|
| $p_{ij} = h_i(B_j)$ | the profit (hit rate of cache $i$) |
| $w_{ij} = bram_i(B_j)$ | the cost (block RAM consumption of cache $i$) |
| $C$ | the global constraint on the available block RAM resources |
| $x_{ij} \in \{0, 1\}$ | a binary variable, |

where $i = 0 \dots K-1$ iterates over caches and $j = 0 \dots N-1$ iterates of cache sizes. We phrase the maximisation problem as a *Multiple-Choice Knapsack Problem* (MCKP) [14] as follows:

$$\begin{aligned} &\text{maximise } \textstyle\sum_{i=0}^{K-1} \sum_{j=0}^{N-1} p_{ij} x_{ij} \\ &\text{subject to } \textstyle\sum_{i=0}^{K-1} \sum_{j=0}^{N-1} w_{ij} x_{ij} \leq C \\ &\text{and } \quad \textstyle\sum_{j=0}^{N-1} x_{ij} = 1, \quad i = 0 \dots K-1 \end{aligned} \tag{5.12}$$

The objective in (5.12) maximises the aggregate hit rate of $K$ caches. The first constraint enforces memory resource limits and the second constraint ensures that, for each cache, exactly one size from the set $\mathcal{B}_i$ is selected by the algorithm. We solve the Knapsack problem with an algorithm by Pisinger et al. [14] based on dynamic programming. The next section describes the code generation and transformations before and after cache sizing.

## 5.5 Experiments

We run our experiments with the four C++ applications from Chap. 4 that traverse, update, allocate and dispose dynamic data structures in heap memory. In contrast to Chap. 4, the parallelised implementations of **Reflect Tree** and **Filter** contain interfaces to shared memory. All applications perform pointer-chasing and are therefore very sensitive to the memory access latency. For brevity, we omit **Tree deletion** benchmark from the previous chapter in this evaluation.

**Merger**. The program builds up four linked lists from scratch performing a sorted insertion of input values, and subsequently merges and disposes the four lists to produce a single sorted output stream. The linked lists are disjoint, the parallelised program does not access shared heap memory as determined by our analysis. Four private scratchpads are inserted in the parallelised implementation.

**Reflect tree**. The application traverses a binary tree and recursively swaps the left and right child pointer of some nodes to produce a partially mirrored tree. The HLS core consists of $P$ parallel units, each of which has two private memory interfaces and one interface to shared memory which holds a running minimum. $P$ coherent scratchpads and a lock service are instantiated for the shared heap region.

**Filter**. This is our running example. The tree, centre sets and linked list data structures are partitioned and supported by private caches and the traversal loop is parallelised. The shared heap-allocated running sum is supported by coherent scratchpads and a lock service.

We use Xilinx Vivado HLS 2014.1 as a back-end C-to-FPGA tool. As of writing of this thesis, LEAP supports Altera FPGA boards as well as several boards with Xilinx FPGAs (Nallatech ACP, XUPV5, HTG-V5, ML605, VC707). Recently, support has been added for Xilinx VC709 boards with two board-level DDR3 memory modules. Here, we implement our benchmarks on a VC707 evaluation board (Virtex 7 FPGA, xc7vx485tffg1761-2, 1GB on-board DDR3 SDRAM). We build the Bluespec-based LEAP framework with Bluespec 2014-07-A. The generated RTL code is integrated into the framework with Bluespec's `import BVI` statement. The complete FPGA designs are implemented in a hybrid flow with Synopsys Synplify Premier 2014.03.1 for logic synthesis and Xilinx Vivado 2014.4 for placement and routing. We report FPGA slices, DSP slices, 36k-BRAMs (18k-blocks count as 0.5 36k-blocks), achieved clock period and total latency (cycle count $\times$ clock period) for the complete FPGA designs (HLS core and multi-cache architecture). The latency is normalised differently depending on the benchmark: latency per input sample for **Merger**, latency per full tree traversal for **Reflect Tree**, and latency per clustering iteration for **Filter**.

We separate this evaluation into two parts: The first part focuses on the performance gained by inserting our multi-cache system and the benefits of specialising it by inserting coherent caches only if necessary. The second part of this evaluation section discusses the automatic scaling of private caches.

### *5.5.1 Hybrid Multi-cache Architectures*

Table 5.1 quantifies the acceleration and resource consumption of parallelisation and the multi-cache architecture $N_c$ is the number of inserted caches. The default size of all caches is 1 kB. For each benchmark, we set the unparallelised ($P = 1$) design

**Table 5.1** Parallelisation and caching (cache size 1 kB)

*P* parallelisation degree; $N_c$ number of caches; *S* speed-up over baseline

| P | $N_c$ | Slices | LUT | FF | DSP | BRAM | Clock rate (MHz) | Latency (ms) | S |
|---|---|---|---|---|---|---|---|---|---|
| **Merger** (250,000 random input key-value pairs) | | | | | | | | | |
| Scratchpads without on-chip caches | | | | | | | | | |
| 1 | 0 | 22,993 | 58,709 | 59,029 | 21 | 571.5 | 100 | 18.0 | **1** |
| 4 | 0 | 24,242 | 67,867 | 67,130 | 19 | 586.5 | 100 | 5.9 | 3.08 |
| Scratchpads with on-chip caches (1 kB) | | | | | | | | | |
| 1 | 1 | 24,885 | 64,860 | 64,820 | 24 | 583.5 | 100 | 19.4 | 0.93 |
| 4 | 8 | 34,184 | 91,401 | 88,830 | 38 | 634.5 | 100 | 6.4 | 2.82 |
| **Reflect Tree** (36,862 tree nodes) | | | | | | | | | |
| Scratchpads without on-chip caches | | | | | | | | | |
| 1 | 0 | 24,944 | 64,471 | 65,953 | 37 | 231.5 | 100 | 547.5 | **1** |
| 2 | 0 | 27,188 | 74,820 | 77,230 | 57 | 248.5 | 100 | 344.9 | 1.59 |
| 4 | 0 | 35,891 | 95,483 | 99,269 | 97 | 360.5 | 100 | 194.0 | 2.82 |
| Scratchpads with on-chip caches (1 kB) | | | | | | | | | |
| 1 | 3 | 27,844 | 70,662 | 72,437 | 46 | 243.5 | 100 | 320.6 | 1.71 |
| 2 | 6 | 32,238 | 88,010 | 89,730 | 73 | 278.5 | 100 | 153.7 | 3.56 |
| 4 | 12 | 43,747 | 118,226 | 123,215 | 129 | 408.5 | 100 | 79.9 | 6.85 |
| **Filter** (32,767 kd-tree nodes, 128 clusters) | | | | | | | | | |
| Scratchpads without on-chip caches | | | | | | | | | |
| 1 | 0 | 26,249 | 72,980 | 74,050 | 57 | 275 | 100 | 897.3 | **1** |
| 2 | 0 | 32,253 | 91,518 | 91,964 | 97 | 347.5 | 100 | 594.5 | 1.51 |
| 4 | 0 | 42,636 | 128,163 | 128,587 | 179 | 486.5 | 100 | 415.8 | 2.16 |
| Scratchpads with on-chip caches (1 kB) | | | | | | | | | |
| 1 | 4 | 29,518 | 83,077 | 83,493 | 67 | 296 | 100 | 464.7 | 1.44 |
| 2 | 8 | 39,284 | 110,179 | 110,419 | 132 | 383.5 | 100 | 240.5 | 3.73 |
| 4 | 16 | 54,736 | 163,849 | 164,620 | 218 | 558.5 | 100 | 145.6 | 6.16 |

with no caches as a baseline reference (top row for each benchmark). The ratio *S* is the speed-up of each configuration compared to the baseline reference case ($S = 1$).

Adding single caches to the unparallelised implementations ($P = 1$) brings a speed-up of 1.71× and 1.44× for **Reflect tree Filter**, respectively. Parallelisation with $P = 4$ results in 2.16× to 2.82× speed-up over the unparallelised baseline if the memory interface is not supported by caches. We observe further latency improvements when these parallelised applications are supported by multiple caches, which provides an overall acceleration of 6.16× to 6.85× for the tree-based benchmark. The small caches mostly reduce the memory access time for the stack and centre set data structures, as opposed to the tree data structures which are substantially larger.

**Table 5.2** Cost increase of all-coherent default compared to application-specific hybrid scratchpad architectures

$P$ parallelisation degree; $N_C$ number of caches

| P | $N_C$ | Slices | DSP | BRAM | Clock period/ns | Latency/ms | Area–time product |
|---|---|---|---|---|---|---|---|
| **Merger** (250,000 random input key-value pairs) | | | | | | | |
| 4 | 8 | 42,875 (25.4%) | 145 (281.6%) | 642 (1.2%) | 10.0 (0.0%) | 7.83 (22.6%) | 335.7 slices · s (53.8%) |
| **Reflect tree** (36,862 tree nodes) | | | | | | | |
| 2 | 6 | 35,683 (10.7%) | 122 (67.1%) | 330 (18.5%) | 10.0 (0.0%) | 220.0 (43.1%) | 7850.1 slices · s (58.4%) |
| 4 | 12 | 52,665 (20.4%) | 220 (70.5%) | 504 (23.4%) | 10.2 (2.2%) | 128.5 (60.7%) | 6765.5 slices · s (93.5%) |
| **Filter** (32,767 kd-tree nodes, 128 clusters) | | | | | | | |
| 2 | 8 | 45,579 (16.0%) | 190 (43.9%) | 367 (−4.3%) | 10.0 (0.0%) | 366.1 (52.2%) | 16687.8 slices · s (76.6%) |
| 4 | 16 | 65,412 (19.5%) | 375 (72.0%) | 644 (15.3%) | 10.1 (0.9%) | 208.2 (43.0%) | 13615.8 slices · s (70.9%) |

As we shall see in Sect. 5.5.4, the system performance is further improved by cache scaling. **Merger** is an extreme case in this evaluation because inserting the small 1 kB caches slightly slows down the implementations for both $P = 1$ and $P = 4$. The reason is the size of the data structures: only 2048 list elements fit in the caches, which is a small fraction of the entire data structure, resulting in a poor hit rate. The improvement of the memory access latency by the caches thus does not outweigh the small overhead in terms of cycle count because of the buffered banked cache memories [9]. Parallelisation improves the net speed-up, but we shall see in the next sections that scaled-up caches further improve the overall latency significantly.

In addition to aggregate latency, we evaluate the benefit of cache architecture specialisation. Our analysis determines that **Merger** requires $P$ private SPs, while **Reflect Tree** and **Filter** require a hybrid architecture consisting of private and coherent SPs. We compare the implementation results of our application-specific architectures to an 'all-coherent' scenario where no knowledge of disjoint heap regions is available to generate the multi-cache system. Firstly, such a scenario requires a commutativity analysis for safe parallelisation for all heap updates which significantly increases the burden of analysis. Secondly, all SPs must be supported with a coherency network by default. We focus on the second aspect here and quantify the additional cost of such an all-coherent architecture in terms of loss of efficiency: Table 5.2 lists the implementation results for the designs with all-coherent SPs. Each row also shows the increase in resource consumption, latency and the slices–latency product of the all-coherent (AC) default compared to the corresponding hybrid (HY) SP architecture in Table 5.1 which uses knowledge of private and shared heap regions ($\frac{AC-HY}{HY}$ in %).

The AC versions use more logic and have longer latencies. The resource overhead is especially noticeable for DSP slices (43.9 up to 281.6%), but is also substantial for logic slices (10.7–25.4%). The area overhead is particularly large for **Merger**, because the application-specific SP architecture does not use a coherency network at all, so the difference is larger. The access latencies due to the additional coherency network are notably longer. Finally, we compare the efficiency of the implementations by the area–time product. For $P = 4$, our disjointness analysis and the ability to instantiate cheap private caches whenever possible brings an overall improvement of the slices–latency product of 53.8–93.5% (69.3% on average).

The results above quantify the advantage of a specialised application-specific multi-cache system. The following sections discuss the validation of the resource and hit rate estimation and the performance improvements and trade-offs by scaling up the private caches in the above hybrid multi-cache system.

### 5.5.2 *Validating the BRAM Estimation for Automated Cache Scaling*

Our automatic cache scaling relies on the ability to estimate the amount of BRAM used by the HLS implementation for core-internal storage. Once the tool decided which variables in the code go into BRAM, a conservative estimate can be easily made. Vivado HLS, for example, provides such an estimate after RTL generation. For brevity, we only show the validation for a parallelisation degree of $P = 2$. Table 5.3 compares the high-level BRAM estimation of 36k-RAM blocks with post placement-and-routing (PAR) results. Additional BRAM is used in FIFOs of the wrappers connecting HLS bus interfaces to LEAP scratchpad ports. Similarly, our scratchpad interfaces (scratchpads without caches) contain some FIFOs as well. The RAM usage of these FIFOs can be precisely determined from the Verilog/Bluespec System Verilog code. The LEAP-based framework uses a fixed amount of RAM. The only uncertainty are the estimates made by the HLS tool, but these are always higher than the post-PAR consumption because of bit truncations made by the RTL synthesis tool. We also include a 10% security margin in the left-over portion that will be used for the cache implementations, i.e. we scale the cache sizes up such that the estimated memory consumption of the HLS core and its interface wrappers, the platform and the caches reaches at most 90% of the BRAM resources available on the chip.

### 5.5.3 *Validating Cache Performance Estimation*

We validate our cache model with measurements of the actual hit/miss rates. LEAP Scratchpads collect the number of hits and misses for each cache during execution of the application. Table 5.4 compares the measured individual hit rate $h_{\text{meas}}$ for different cache sizes with the estimated values $h_{\text{est}}$ from Algorithm 8. The hit rates are

**Table 5.3** High-level BRAM estimation accuracy (results in 36k-RAM blocks)

| Design component | Estimate | Post-PAR |
|---|---|---|
| **Merger** ($P = 4$, 8 scratchpads) | | |
| HLS core | 512 | 512 |
| Interface wrapper | 12 | 12 |
| Scratchpad internal FIFOs | 12 | 12 |
| LEAP platform (without scratchpads, fixed) | 50.5 | 50.5 |
| Total consumption without caches | 586.5 | 586.5 |
| Unused left-over blocks (xc7vx485tffg1761-2) | 340.5 | 340.5 |
| **Reflect tree** ($P = 2$, 6 scratchpads) | | |
| HLS core | 208 | 158 |
| Interface wrapper | 21 | 21 |
| Scratchpad internal FIFOs | 19 | 19 |
| LEAP platform (without scratchpads, fixed) | 50.5 | 50.5 |
| Total consumption without caches | 298.5 | 248.5 |
| Unused left-over blocks (xc7vx485tffg1761-2) | 628.5 | 678.5 |
| **Filter** ($P = 2$, 8 scratchpads) | | |
| HLS core | 275 | 241 |
| Interface wrapper | 32 | 32 |
| Scratchpad internal FIFOs | 24 | 24 |
| LEAP platform (without scratchpads, fixed) | 50.5 | 50.5 |
| Total consumption without caches | 381.5 | 347.5 |
| Unused left-over blocks | 545.5 | 579.5 |

calculated with $h = n_{\text{hit}}/(n_{\text{hit}} + n_{\text{miss}})$ and the cache sizes are given in terms of 64bit lines. We also include the relative error. Additionally, we compare the stack distance-based approximation in [12] ($h_{\text{est}}^{\text{SD}}$, $\text{error}^{\text{SD}}$) with our estimator. For brevity, we show results only for two private caches of the **Reflect tree** benchmark and for $P = 2$. Our estimation matches exactly the measured hit/miss counts, i.e. Algorithm 8 models our direct-mapped caches perfectly. The approximation by Brehob and Enbody [12] tends to underestimate the hit rate of direct-mapped caches, an observation also made in [12].

### 5.5.4 *Latency and Resource Utilisation After Custom Cache Scaling*

Our technique improves the aggregate hit rate of the multi-cache architecture. The following results show the impact of the custom cache sizing on the overall execution latency and on the FPGA resource usage once we scale the hybrid multi-cache sys-

**Table 5.4** Cache hit/miss count estimation for two private caches in **Reflect tree**

| Cache size | $h_{meas}$ (%) | $h_{est}$ (%) | Error (%) | $h_{est}^{SD}$ (%) | Error$^{SD}$ (%) |
|---|---|---|---|---|---|
| **Cache 0** | | | | | |
| 1024 | 86.46 | 86.46 | 0.00 | 84.71 | −2.06 |
| 8192 | 87.44 | 87.44 | 0.00 | 87.12 | −0.36 |
| 32,768 | 87.49 | 87.49 | 0.00 | 87.42 | −0.08 |
| 65,536 | 87.50 | 87.50 | 0.00 | 87.87 | 0.42 |
| 262,144 | 95.83 | 95.83 | 0.00 | 91.43 | −4.82 |
| **Cache 1** | | | | | |
| 1024 | 86.35 | 86.35 | 0.00 | 84.44 | −2.26 |
| 8192 | 87.03 | 87.03 | 0.00 | 87.12 | 0.11 |
| 32,768 | 95.68 | 95.68 | 0.00 | 91.04 | −5.09 |
| 65,536 | 95.68 | 95.68 | 0.00 | 92.94 | −2.94 |
| 262,144 | 95.68 | 95.68 | 0.00 | 94.90 | −0.82 |

tems. All results are obtained from a physical implementation on the VC707 board. For ease of comparison, we include the uncached case and the case with small default size caches (both from Sect. 5.5.1). We compare four cases:

**case 1**. An implementation without any caches (as in Table 5.1)

**case 2**. An implementation with a small fixed cache size of 1024 lines (as in Table 5.1)

**case 3**. An implementation with a fixed size for all caches but scaled up to the maximum possible size

**case 4**. A variably-sized multi-cache system as delivered by our technique in Sect. 5.4

The clock frequency target is set to 100 MHz in all cases and all designs meet this clock constraint. All caches have a line width of 64 bits. Table 5.5 shows the timing as well as the utilisation of LUTs, FFs, DSP slices and 36k-RAM blocks. We also show the aggregate hit rate (measured) of all private caches and the execution latency. We compare the speed-up $S$ with respect to the base case in Sect. 5.5.1 ($P = 1$, no caches).

In addition to more BRAM, we observe a sudden increase in LUT, FF and DSP utilisation once caches are included in the scratchpads. LUTs and FFs increase only marginally when scaling the caches up, leaving the BRAM usage as the limiting factor. The hit rate and latency improvements for **Merger** are substantial and grow steadily with larger cache sizes. There is a significant asymmetry between the linked lists in the application and the large improvement of the variable sizing over a fixed sizing (Cases 3 and 4) is due to the fact that larger caches support longer lists. The overall speed-up after parallelisation, private cache insertion and custom cache sizing is $S = 15.22$ over the baseline.

**Table 5.5** Latency and resource utilisation after custom cache scaling

*S* speed-up over unparallelised, uncached baseline in Table 5.1

| P | Case | LUT | FF | DSP | BRAM | Hit rate | Latency (ms) | *S* |
|---|---|---|---|---|---|---|---|---|
| **Merger** (250,000 random input values, baseline latency: 18.0 ms) | | | | | | | | |
| 4 | 1 | 67,867 | 67,130 | 19 | 586.5 | 0 | 5.9 | 3.08 |
| | 2 | 91,401 | 88,830 | 38 | 634.5 | 5.31% | 6.4 | 2.82 |
| | 3 | 93,528 | 89,184 | 38 | 858.5 | 79.24% | 2.4 | 7.39 |
| | 4 | 92,871 | 89,064 | 39 | 874.5 | 99.12% | 1.2 | 15.22 |
| **Reflect tree** (36,863 tree nodes, baseline latency: 547.5 ms) | | | | | | | | |
| 2 | 1 | 74,820 | 77,230 | 57 | 248.5 | 0 | 344.9 | 1.59 |
| | 2 | 88,010 | 89,730 | 73 | 278.5 | 90.36% | 153.7 | 3.56 |
| | 3 | 88,204 | 89,989 | 73 | 862.5 | 95.69% | 138.4 | 3.95 |
| | 4 | 88,193 | 89,855 | 73 | 944.5 | 99.97% | 112.4 | 4.87 |
| 4 | 1 | 95,483 | 99,269 | 97 | 360.5 | 0 | 194.0 | 2.82 |
| | 2 | 118,226 | 123,215 | 129 | 408.5 | 90.12% | 79.9 | 6.85 |
| | 3 | 136,087 | 125,046 | 126 | 743.5 | 95.50% | 68.7 | 7.97 |
| | 4 | 119,284 | 123,253 | 128 | 736.5 | 98.27% | 57.4 | 9.54 |
| **Filter** (32,767 kd-tree nodes, 128 clusters, baseline latency: 897.3 ms) | | | | | | | | |
| 2 | 1 | 91,518 | 91,964 | 97 | 347.5 | 0 | 594.5 | 1.51 |
| | 2 | 110,179 | 110,419 | 132 | 383.5 | 93.52% | 240.5 | 3.73 |
| | 3 | 111,459 | 110,806 | 116 | 807.5 | 95.95% | 234.9 | 3.82 |
| | 4 | 110,423 | 110,448 | 117 | 711.5 | 98.76% | 229.3 | 3.91 |
| 4 | 1 | 128,163 | 128,587 | 179 | 486.5 | 0 | 415.8 | 2.16 |
| | 2 | 163,849 | 164,620 | 218 | 558.5 | 94.12% | 145.6 | 6.16 |
| | 3 | 168,416 | 165,278 | 221 | 886.5 | 96.19% | 140.4 | 6.39 |
| | 4 | 164,818 | 164,168 | 219 | 878.5 | 98.72% | 138.2 | 6.49 |

For the tree-based benchmarks, we see a different characteristic of the latency improvement. Even small caches lift the aggregate hit rate above 90%. This reflects the behaviour in Fig. 5.4: the stack data structures are very small (but heavily accessed) compared to the tree structure in the average case and a small cache is sufficient to keep all data on-chip. Consequently, the optimisation algorithm in Sect. 5.4.3 opts to use more memory resources for the large tree structure. For **Reflect tree**, this improves the aggregate hit rate by 3–4% compared to a homogeneous maximum sizing. Although the hit rates for **Filter** and **Reflect tree** are similar, the latency improvement from cache scaling for **Filter** is small. This is mainly due to high core-internal computation between memory accesses, which makes the effect of a shorter access time to the tree data less significant. The overall improvement execution time after parallelisation with $P = 4$, hybrid cache insertion and custom cache scaling is 9.54× and 6.49× over the unparallelised and uncached baseline implementation for **Reflect tree** and **Filter**, respectively.

### 5.5.5 *Energy Consumption*

We quantify the impact of our cache insertion and scaling on the overall energy consumption. To this end, we measure the instantaneous power consumption of the FPGA and the board-level SDRAM while the applications are running. We collect power figures for three out of the 12 power rails on the VC707 board: `VCCINTFPGA` is the main supply of the FPGA and `VCCBRAM` is an additional block RAM supply. We combine both to obtain the main supply of the FPGA. The third rail is `VCC1V5`, a supply of the SDRAM. No other rail notably changes its power levels during execution of our applications. We integrate power over the three latencies defined in the previous section; we show the energy per input value for **Merger**, the energy per completed tree traversal for **Reflect tree** and the energy per clustering iteration for **Filter**. Table 5.6 shows the main energy consumption of the FPGA ($E_{\mathrm{FPGA}}$), the energy attributed to the SDRAM ($E_{\mathrm{SDRAM}}$) and the total energy for the four cases

**Table 5.6** Power and energy measurements

$R$: energy reduction compared to Case 1

| P | Case | $P_{\mathrm{FPGA}}$/W | $P_{\mathrm{SDRAM}}$/W | $E_{\mathrm{FPGA}}$/mJ | $E_{\mathrm{SDRAM}}$/mJ | $E_{\mathrm{total}}$/mJ | $R$ |
|---|---|---|---|---|---|---|---|
| **Merger** (250,000 random input values) | | | | | | | |
| 4 | 1 | 1.78 | 1.11 | 10.40 | 6.40 | 16.88 | **1** |
| | 2 | 2.13 | 1.09 | 13.61 | 6.99 | 20.60 | 0.82 |
| | 3 | 2.58 | 1.05 | 6.30 | 2.55 | 8.85 | 1.91 |
| | 4 | 2.57 | 1.01 | 3.05 | 1.19 | 4.24 | 3.99 |
| **Reflect tree** (36,863 tree nodes) | | | | | | | |
| 2 | 1 | 1.85 | 1.16 | 638.23 | 401.11 | 1039.34 | **1** |
| | 2 | 2.00 | 1.16 | 307.51 | 177.68 | 485.19 | 2.14 |
| | 3 | 3.38 | 1.07 | 467.59 | 148.23 | 615.82 | 1.69 |
| | 4 | 3.73 | 1.00 | 419.01 | 112.23 | 531.24 | 1.96 |
| 4 | 1 | 2.13 | 1.15 | 412.30 | 222.46 | 634.76 | **1** |
| | 2 | 2.34 | 1.04 | 186.68 | 83.24 | 269.92 | 2.35 |
| | 3 | 3.06 | 1.07 | 210.40 | 73.37 | 283.77 | 2.24 |
| | 4 | 3.26 | 1.14 | 187.09 | 65.56 | 252.64 | 2.51 |
| **Filter** (32,768 kd-tree nodes, 128 clusters) | | | | | | | |
| 2 | 1 | 1.96 | 1.16 | 1166.57 | 691.28 | 1857.85 | **1** |
| | 2 | 2.15 | 1.01 | 517.59 | 243.88 | 761.47 | 2.71 |
| | 3 | 3.16 | 1.03 | 742.82 | 241.43 | 984.25 | 1.89 |
| | 4 | 2.93 | 1.02 | 671.32 | 234.77 | 906.09 | 2.05 |
| 4 | 1 | 2.25 | 1.31 | 936.46 | 542.55 | 1479.01 | **1** |
| | 2 | 2.77 | 1.05 | 402.50 | 152.94 | 555.44 | 2.66 |
| | 3 | 3.27 | 1.03 | 459.55 | 144.19 | 603.74 | 2.45 |
| | 4 | 3.53 | 1.08 | 488.11 | 148.73 | 636.84 | 2.32 |

above. We also show the energy improvement $R$ compared to Case 1 (uncached parallel implementation). The instantaneous power consumption is steady during the execution, so Table 5.6 also shows the mean power consumptions $P_{\mathrm{FPGA}}$ and $P_{\mathrm{SDRAM}}$.

Including caches always comes along with an increased power consumption of the FPGA. For large caches, the extra power consumption is significant (up to 102%). The latency reduction must be large enough to counter this effect and improve $E_{\mathrm{FPGA}}$ and $E_{\mathrm{total}}$. Large caches always improve the energy consumption with respect to a cache-less memory interface in our implementations. In all benchmarks, the application-specific cache sizing outperforms fixed sizing in terms of energy reduction. Interestingly, small caches (Case 2) in the **Reflect tree** and **Filter** benchmark ($P = 2$) have the best performance in terms of energy. The trade-offs when optimising for energy instead of hit rate are different because the increased power consumption of large caches is not always compensated by the reduction of execution time.

## 5.6 Summary

Mapping dynamic memory operations to FPGAs is difficult, both in terms of analysis and implementation. In this chapter, we present an HLS design aid for synthesising pointer-based C/C++ programs into efficient FPGA applications. We target applications that perform computation on large heap-allocated data structures and that require access to an off-chip memory. We leverage and extend the separation logic-based static program analysis in Chap. 4 to determine whether different program parts access disjoint, non-overlapping regions in the monolithic heap space in which case we trigger automated source-to-source transformations that automatically parallelise the application. Our extended analyser also detects heap regions that are shared by multiple accessors in the parallelised implementation. An additional commutativity analysis decides whether the parallelisation in the presence of shared memory regions is semantics-preserving. The information provided by the heap analyses is used to optimise the interface between the parallelised HLS kernel and an off-chip memory: we generate an application-specific multi-cache architecture where disjoint heap partitions are mirrored in private, independent on-chip caches and interfaces to shared heap regions are supported where necessary with on-chip caches backed by (inherently more expensive) coherency mechanisms and a synchronisation service.

In our experiments with three heap-manipulating C++ benchmark applications, we observe a speed-up of up to 6.9× after parallelisation and generation of a multi-scratchpad architecture compared to the unparallelised application and uncached access to the off-chip memory. We also quantify the benefit of extracting application-specific knowledge about disjoint and shared heap memory regions: our hybrid multi-scratchpad architecture consisting of private and coherent scratchpads outperforms a default all-coherent version by 69.3% on average in terms of the area–time product.

We extend the hybrid cache synthesis from generating caches with a default size to custom cache sizing. The add-on automatically uses up the left-over BRAM to scale up the size of the private on-chip caches. Secondly, the size of each cache is set individually in order to reach a size distribution across the parallel caches that maximises the aggregate hit rate of the multi-cache architecture. The pre-synthesis cache performance estimation is based on a high-level cache model and on the memory reference trace of the application obtained from automated profiling. We cast the cache size assignment into a Multiple-Choice Knapsack Problem to find the best size distribution for a given reference trace.

We evaluate the left-over BRAM and cache hit rate estimation, and we demonstrate the latency improvements obtained from our technique using three benchmarks with irregular memory access patterns running on a VC707 FPGA board. We observe up to a 4.9× speed-up compared to a cacheless memory interface when scaling each on-chip cache to the same maximal size. Our variably-sized multi-cache system also delivers up to a 2.1× latency improvement (1.3× on average) compared to the one-size-fits-all solution. The overall reduction of execution time after parallelisation with $P = 4$, insertion of the hybrid multi-cache system and custom cache scaling is up to 15.2× (9.8× on average) over an unparallelised and uncached implementation. Although the insertion of large on-chip caches has a significant impact on the power consumption of the FPGA, we show that our variably-sized multi-cache configuration reduces the total energy by 2.5× (on average) compared to a cacheless memory interface. The work discussed in this chapter was first published in [15, 16].

## References


1. M. Adler, K.E. Fleming, A. Parashar, M. Pellauer, J. Emer, Leap scratchpads: automatic memory and cache management for reconfigurable logic, in *Proceedings of the International Symposium on Field Programmable Gate Arrays (FPGA)* (2011), pp. 25–28
2. H.-J. Yang, K. Fleming, M. Adler, J. Emer, LEAP shared memories: automating the construction of FPGA coherent memories, in *Proceedings of the IEEE International Symposium on Field-Programmable Custom Computing Machines (FCCM)* (2014), pp. 117–124
3. K. Fleming, H.-J. Yang, M. Adler, J. Emer, The LEAP FPGA operating system, in *Proceedings of the International Symposium on Field Programmable Logic and Applications (FPL)* (2014), pp. 1–8
4. M.C. Rinard, P.C. Diniz, Commutativity analysis: a new analysis technique for parallelizing compilers. ACM Trans. Program. Lang. Syst. **19**(6), 942–991 (1997)
5. Z3: An Efficient SMT Solver. http://z3.codeplex.com/documentation/. Accessed 04 Sep 2014
6. A. Charlesworth, The undecidability of associativity and commutativity analysis. ACM Trans. Program. Lang. Syst. **24**(5), 554–565 (2002)
7. E.S. Chung, J.C. Hoe, K. Mai, CoRAM: an in-fabric memory architecture for FPGA-based computing, in *Proceedings of the ACM/SIGDA International Symposium on Field Programmable Gate Arrays* (2011), pp. 97–106
8. B. Cook, A. Gupta, S. Magill, A. Rybalchenko, J. Simsa, S. Singh, V. Vafeiadis, *Finding Heap-Bounds for Hardware Synthesis, in Formal Methods in Computer-Aided Design* (IEEE, New York, 2009), pp. 205–212
9. H.-J. Yang, K. Fleming, M. Adler, F. Winterstein, J. Emer, Scavenger: automating the construction of application-optimized memory hierarchies, in *Proceedings of the International Conference on Field Programmable Logic and Applications (FPL)* (2015), pp. 1–8
10. E.G. Coffman, P.J. Denning, *Operating Systems Theory* (Prentice Hall, Englewood Cliffs, 1973)

11. M. Hill, A. Smith, Evaluating associativity in CPU caches. IEEE Trans. Comput. **38**(12), 1612–1630 (1989)
12. M. Brehob, R. Enbody, *An Analytical Model of Locality and Caching Department of Computer Science*, Michigan State University, Technical report (1996)
13. K. Beyls, E.H. DHollander, Reuse distance as a metric for cache behavior, in *Proceedings of the IASTED Conference on Parallel and Distributed Computing and Systems* (2001), pp. 617–662
14. D. Pisinger, A minimal algorithm for the 0–1 Knapsack problem. Oper. Res. **45**, 758–767 (1994)
15. F. Winterstein, K. Fleming, H.-J. Yang, J. Wickerson, G. Constantinides, Custom-sized caches in application-specific memory hierarchies, in *Proceedings of the International Conference on Field Programmable Technology (ICFPT)* (2015), pp. 144–151
16. F. Winterstein, K. Fleming, H.-J. Yang, S. Bayliss, G. Constantinides, MATCHUP: memory abstractions for heap manipulating programs, in *Proceedings of the ACM/SIGDA International Symposium on Field-Programmable Gate Arrays (FPGA)* (2015), pp. 136–145

# Chapter 6
# Conclusion

This thesis extends the scope of high-level synthesis to efficient hardware implementations from heap-manipulating programs. This research direction is motivated by the fact that hardware synthesis and design optimisations for heap-manipulating code are beyond the scope of state-of-the-art HLS tools and most HLS techniques to date. We underpin this motivation with a case study in Chap. 2 which compares the performance gap between HLS and hand-written RTL implementations of a data flow-centric $K$-means clustering algorithm and an algorithm for the same problem that uses dynamic memory management and is based on the traversal of a pointer-linked tree data structure. Our results show that both a carefully designed RTL and HLS implementation of the latter algorithm results in faster and more efficient hardware implementations. We furthermore quantify the benefits of hardware implementations of a sophisticated algorithm that uses structured data, organised in dynamically allocated, irregular, pointer-linked data structures.

The direct HLS implementation of the pointer-based filtering algorithm requires code transformations to enable synthesisability. Furthermore, the latency is initially degraded by 26.6× compared to the hand-crafted RTL implementation. We narrow this significant performance gap and improve the former latency by 8× with source code transformations that partition and privatise data structures accessed through pointers to enable parallelisation and pipelining of the loop traversing the pointer-linked data structure. Our case study exposes the lack of support for effective design automation optimisations for codes containing heap-allocated data structures and pointer chasing in the tool under test which we consider a representative example of current generation HLS tools. However, these limitations can be removed with extensive code refactoring that draws on memory disambiguation and data dependence information, which is not provided in current state-of-the-art HLS flows if the memory accesses are made through heap-directed pointers.

In Chap. 4, we address the automation of two key code optimisations. The goal is to automate the memory partitioning and semantics-preserving parallelisation, which were performed manually in our case study. We develop an automated analysis of



F. Winterstein, *Separation Logic for High-level Synthesis*, Springer Thesis,
DOI 10.1007/978-3-319-53222-6_6

data dependencies carried by pointer-based data structures, which identifies disjoint regions in the monolithic heap memory space. Chapter 4 contains the key enabling research contribution in this thesis: the application and extension of a static program analysis framework based on separation logic, a logic for efficient reasoning about programs that dynamically allocate and dispose memory space and access data in heap memory. Separation logic arose in the context of formal software verification and recently made its way into commercial verification tools [1].

The potential of separation logic, which has made it a widely used framework in the verification domain, remains largely unexplored in an HLS context to date. This thesis provides a deep investigation of separation logic-based program analyses for code optimisations in hardware compilers. We show how existing techniques in the separation logic framework, symbolic execution and loop invariant synthesis, which were originally developed in a verification context, can be modified and extended so as to repurpose the analysis for ruling out heap-carried data dependencies between different execution phases of a program. We also extend an existing approach to a heap footprint analysis, which allows us to partition the monolithic heap space into disjoint fragments (heaplets), a prerequisite for distributing heap-allocated data across physically separated memory banks. A key advantage of our technique is its ability to handle `while`-loops with data-dependent loop condition, enclosed data-dependent conditionals and unknown iteration count. This feature and the ability to reason about dynamically allocated data structures distinguishes our analysis from the polyhedral model, the most powerful and widely used loop optimisation framework to date.

The information provided by the static program analysis, the legality of parallelisation (delivered by proving the absence of heap-induced data dependencies) and an assignment of heaplets to on-chip memory partitions, is used by automated source-to-source transformations that ensure the synthesis of parallel loop kernels and parallel banks of on-chip memory. The distribution of data across memory banks is specific to the application and is guided by the assignment of heap partitions to memories delivered by our analysis, resulting in a specialised on-chip memory architecture. The fact that our analysis can prove the absence of communication between the parallel loop kernels and the disjointness of the heap portions mapped to different memory banks allows us to avoid the synthesis of unnecessary interconnects and synchronisation hardware between parallel functional units. To the best of our knowledge, this work is the first application of separation logic in the context of an automated memory optimisation tool flow for HLS. In Sect. 4.4, we evaluate our technique with Xilinx Vivado HLS as an exemplary state-of-the-art HLS tool for FPGAs. Using several pointer-chasing benchmarks, we firstly show that the use of native optimisation directives of the tool does not result in physical heap memory partitioning and parallelisation. Secondly, we show that the HLS implementations parallelised by our tool achieve the expected acceleration by a factor of $1.8\times - 5.3\times$ compared to the direct HLS implementations.

In Chap. 5, we remove the restriction that the application data must fit in on-chip memory and that the heap-allocated data structures must be fully partitionable into disjoint portions to trigger parallelisation. We address the first restriction by synthesising memory architectures which, by default, place all heap-allocated data in a large off-chip memory and insert a parallel on-chip multi-cache system which mirrors parts of the off-chip data and provides fast data access. We then extend the program analysis from Chap. 4 to allow sharing: parallelising source transformations are triggered if some parts of the data structures in the heap can be partitioned into disjoint portions and the data structures for which this disjointness proof fails are marked as shared. We include an additional commutativity analysis to determine whether the parallelisation in the presence of shared memory regions is legal.

We use the disjointness/sharing information to automatically generate cheap private, independent on-chip caches whenever possible and, for shared memory access, to instantiate more expensive caches that are backed by a coherency network and a lock-based synchronisation. Our results show a speed-up of up to 6.9× after parallelisation and cache insertion compared to the unparallelised application and cache-less memory interfaces. Without the assistance of our disjointness/sharing analysis, a CAD flow may instantiate coherent caches by default. We compare this solution with our hybrid cache system and show that the specialisation of the multi-cache architecture, enabled by our static analysis, results in an average improvement of the overall area-time product (and hence architecture efficiency) by 69.3% across our benchmarks.

In addition to automatically deciding about the type of the synthesised caches, we add automatic cache scaling using spare on-chip memory resources in an extension of the cache synthesis flow in Sect. 5.4. We use pre-synthesis profiling information to predict the hit rate of private caches in the memory hierarchy. Each private cache is then assigned an individual size and the cache sizes are automatically distributed across the multi-cache system such that the predicted aggregate hit rate is maximised. Compared to the small default size for all caches as in Sect. 5.3, the variably-sized, scaled multi-cache architecture improves execution time by 1.6× on average.

In summary, the hardware implementations that arise from our pointer-chasing HLS benchmarks, which were parallelised with the technique in Chap. 4 and connected to a cacheless off-chip memory interface, run 2.1× faster (on average) than the direct HLS implementations. This gain is due to the fact that the HLS tool does not detect the parallelisation opportunity if the source code is not altered prior to RTL generation. The insertion of the application-specific multi-cache system from Chap. 5, utilising private caches with custom scaling whenever possible, further accelerates these implementations, resulting in an end-to-end gain of 7.1× speed-up on average and up to 15.2× speed-up for a particular application.

We end this summary with a discussion of some directions for future research. The planned extensions of the program analysis and code generation infrastructure developed in this thesis target both the performance of the analysis and the extension of the scope of our technique to further applications.

## 6.1 Outlook

**Higher parallelisation degrees**. Future work will extend the analysis to scale the designs to higher parallelisation degrees, which requires our analysis to decide between several alternatives as to how the heap can be partitioned. Our current cut-point insertion in Sect. 4.2 greedily searches for an initial partitioning solution and then attempts to prove its validity. If we were to parallelise the motivating example in Sect. 4.1 with a higher degree, several valid alternatives for the cut-point assignment would arise and our analysis would choose the first one (splitting the left sub-tree twice instead of splitting each left and right sub-tree once). However, without further guidance, it cannot guarantee that the work load is distributed uniformly across parallel workers and hence the selected alternative results in the best acceleration. In the experiment section in Chap. 5 we scale our tree-based benchmarks to a parallelisation degree of four and manually guide the analysis to select the best cut-point assignment. Future work can address the automation of fair work load distribution either by including the ability of comparing valid partitioning alternatives in the analysis or by automatically synthesising an additional run-time load balancing network such as that proposed by Ramanathan et al. [2].

**Inferring recursive heap predicates**. The ability of our technique to analyse `while`-loops with data-dependent iteration count relies on the convergence of the fix-point calculation as discussed in Sect. 4.2.2. The fix-point calculation uses a predefined set of recursive predicates for common data structures such as trees, lists, lists with additional pointers to singleton heaplets and sub-trees, which allow us to cover a large range of pointer-based programs. However, we may find applications which use more exotic data structures for which no heap predicate in our current set applies. Future work will address the integration of techniques for automatic inference of recursive heap predicates, such as the work by Guo et al. [3], which further broadens the applicability of our heap analyser.

**Extending the code support**. The fact that our analysis and code transformations are made at the level of LLVM IR allows the seamless integration of our technique into many state-of-the-art HLS flows. The analysis builds an internal representation of the program from the LLVM IR as described in Sect. 4.3.1. Our tool supports a range of `Clang`-generated LLVM IR codes. However, the translation current into the internal representation does not support full-featured LLVM code. For example, instructions related to exception handling are not supported. More importantly, our current analysis requires that all sub-functions in the loop-under-analysis that contain heap-manipulating code be 'inlined'. Future work will extend the coverage to full-featured LLVM code and include the analysis of a call graph. The latter can be addressed with a compositional bottom-up analysis which computes procedure summaries by inferring the pre- and post-condition specifications of sub-functions in separation logic and which can be implemented using a technique called bi-abduction [4]. We plan to adopt the compositional approach using bi-abduction in future work.

**Pipelining**. Our analysis identifies disjoint heap regions accessed by the program to rule out data dependencies and to enable spatial parallelisation. The scope of this dependence analysis can be extended to promote the automatic construction of pipelined data paths from loops. HLS tools allow the user to construct custom pipelines in hardware, where the tool relies on a dependence analysis to decide in which intervals subsequent loop iterations can be scheduled for execution in the pipeline. As we discuss in Chap. 2, the lack of the ability to reason about heap-carried dependencies prevents current tools from pipelining loops traversing heap-allocated data structures. Furthermore, additional loop transformations, such as loop flattening and loop distribution, may be required to expose the possibility of efficient pipelining to an HLS tool as described in Sect. 2.4.2. Future work in the area comes in two parts: Firstly, the analysis will be extended to mark the absence and potential presence of dependencies between loop iterations. This extension requires including a notion of time in our analysis which we plan to address with a combination of separation logic formulae and temporal operators. Secondly, we plan to extend our code transformation framework to support loop transformations and to generate application-specific pipeline circuitry such as in [5].

**Prefetching**. The automatic cache insertion in Chap. 5 can be further extended by synthesising application-specific prefetching units which fetch data from off-chip memory and store it in the on-chip caches in advance. Prefetching in microprocessors is usually based on 'learning' memory access patterns while the program executes and then speculatively prefetching data accordingly. In custom hardware implementation, a tool can build a specialised prefetcher which can prefetch data more accurately. In order to enable the automated synthesis of such units, we plan to extend our heap footprint analysis to provide information about when data is available for prefetching and when it is used by the application. The approach to describe when heap-allocated data is used will also be based on the combination of separation logic formulae and temporal operators.

**Recursion**. The tree-based benchmarks in Chaps. 4 and 5 implement the tree traversal with a `while`-loop and an explicit stack. Programmers may opt to write such programs using recursion instead of explicitly describing a stack. However, similar to dynamic memory allocation, synthesisable recursion is a feature missing from all common HLS flows. Recent work [6] presents a canonical scheme for translating recursive programs into synthesisable code by automatically constructing a stack and the required control flow. We believe that our HLS design aid for memory hierarchy synthesis, memory partitioning and parallelisation is a natural fit for such a translator in that our tool can be added as a back-end to it. Future work will explore the automatic optimisation of recursive programs in an HLS context using the techniques developed in this thesis.

**Modelling coherency networks**. The application-specific multi-cache system can be further extended by including cache size scaling for coherent caches. The current framework in Sect. 5.4 only models and scales private, independent caches. Future work will focus on a model of the coherency protocol in a cache architecture consisting of coherent caches. The cache hit rate estimation of such a coherent cache network must, in addition to cold, conflict and capacity misses, take additional invalidation and owner misses due to interfering accesses by other caches into account. An accurate prediction of this new class of cache misses relies on the knowledge of the exact interleaving of memory accesses by the parallel units, which is contradictory to our approach of modelling the cache performance before RTL generation.

**Modelling energy consumption**. Our current cache sizing framework aims at hit rate maximisation. The majority of our experiments in Sect. 5.5.5 show that maximally scaled caches also result in the lowest overall energy consumption because the run-time reduction outweighs the increased on-chip power consumption due to cache insertion. However, in one benchmark we observe that small caches are more beneficial than large caches in terms of energy consumption, which suggests that the optimal cache sizing changes when we optimise for energy instead of aggregate hit rate. Future work will address the development of an energy model that can be used to minimize the energy consumption of our multi-cache system.

## 6.2 Final Remarks

This thesis offers a program analysis and code transformation infrastructure that enables parallelisation and hardware-specific optimisations of the memory subsystem for the synthesis of efficient hardware implementations from heap-manipulating codes, extending the scope of state-of-the-art HLS tools to this type of programs. We view this extension as an important step towards the support of full-featured C/C++ code in future HLS flows and we envisage pointer-based codes gaining importance in future hardware designs. As a particular example, we believe that the techniques developed in this thesis are useful in future programming environments that target tightly coupled microprocessor-FPGA systems which have great potential to arise in data centre applications [7] in the future. In particular, the C-based OpenCL 2.0 standard [8] that targets such programming environments allows developers to write accelerator codes that directly share dynamically allocated pointer-linked data structures with the software executing on the host microprocessor. The automated analyses developed in this thesis enable efficient implementations of accelerator kernels that process such data structures and thus have potential to improve the programmability and efficient use of these hybrid systems.

## References

1. C. Calcagno, D. Distefano, Infer: an automatic program verifier for memory safety of C programs, in *Proceedings of the International Conference on NASA Formal Methods* (Springer, Heidelberg, 2011), pp. 459–465

2. N. Ramanathan, J. Wickerson, F. Winterstein, G.A. Constantinides, A case for work-stealing on FPGAs with OpenCL atomics, in *Proceedings of the ACM/SIGDA International Symposium on Field-Programmable Gate Arrays (FPGA)* (2016), pp. 48–53
3. B. Guo, N. Vachharajani, D.I. August, Shape analysis with inductive recursion synthesis. ACM SIGPLAN Notices **42**(6), 256 (2007)
4. C. Calcagno, D. Distefano, P. O'Hearn, H. Yang, Compositional shape analysis by means of Bi-abduction. ACM SIGPLAN Notices **44**(1), 289–300 (2009)
5. A. Morvan, S. Derrien, P. Quinton, Efficient nested loop pipelining in high level synthesis using polyhedral bubble insertion, in *Proceedings of the International Conference on Field-Programmable Technology* (2011), pp. 1–10
6. D.B. Thomas, Synthesisable recursion for C++ HLS tools, in *Proceedings of the IEEE International Symposium on Field-Programmable Custom Computing Machines (FCCM)* (2016)
7. A. Putnam, A. Caulfield, E. Chung, D. Chiou, K. Constantinides, J. Demme, H. Esmaeilzadeh, J. Fowers, G. Gopal, J. Gray, M. Haselman, S. Hauck, S. Heil, A. Hormati, J.-Y. Kim, S. Lanka, J. Larus, E. Peterson, S. Pope, A. Smith, J. Thong, P. Xiao, D. Burger, A reconfigurable fabric for accelerating large-scale datacenter services, in *Proceedings of the ACM/IEEE International Symposium on Computer Architecture (ISCA)* (2014), pp. 13–24
8. The OpenCL Specification, Version 2.0. https://www.khronos.org/registry/cl/specs/opencl-2.0.pdf. Accessed 19 Jan 2016

# Appendix A
# Context-Aware Heap Analysis

This section extends the heap analysis of Sect. 4.2 to a *context-aware* analysis. As above, we use a running example to explain our approach. Listing A.1 shows the function reflectTree which traverses a heap-allocated tree structure. The traversal is not destructive as opposed to the motivating example in Listing 4.1. The auxiliary functions push and pop are equivalent to the ones in Listing 4.1. We first explain the problem that occurs when extending our heap analysis to the context.

```
1  //main traversal function
2  void reflectTree(treeNode *root) {
3    // loop preamble
4    stackRecord *s = push(root, NULL);
5    // loop-under-test
6    while (s != NULL) {
7      treeNode *u;
8      s = pop(&u, s);
9      treeNode *l = u->left;
10     treeNode *r = u->right;
11     u->left = r;
12     u->right = l;
13     if (u->left!=NULL && u->right!=NULL)
       {
14       s = push(u->right, s);
15       s = push(u->left, s);
16     }
17   }
18 }
```

Listing A.1 C-like pseudo code for a tree reflection.



F. Winterstein, *Separation Logic for High-level Synthesis*, Springer Thesis,
DOI 10.1007/978-3-319-53222-6

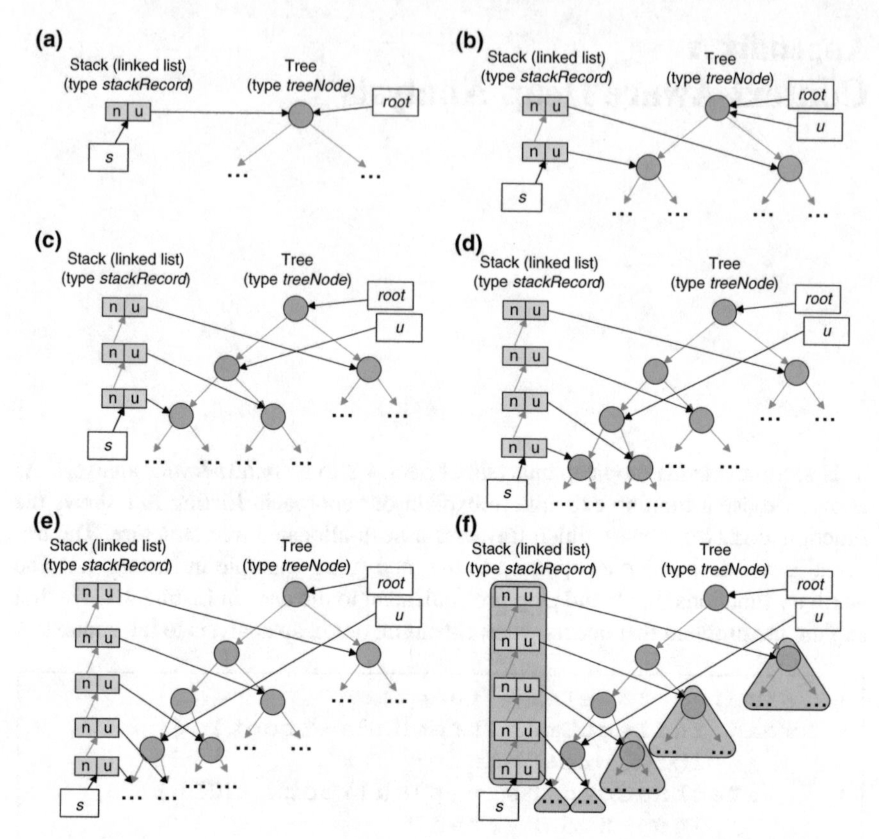


**Fig. A.1** Heap layout and pointer assignment during the first four iterations of the loop in Listing A.1

The program state right before entering the loop in Line 6 is:

$$s \mapsto [\texttt{u} : root, \texttt{n} : \texttt{nil}] * tree(root) \tag{A.1}$$

Figure A.1 shows snapshots of the heap-allocated data structures and pointer assignment during the execution of four iterations of the loop. Our symbolic execution analyses all control flow paths, but here we show a specific path only: Fig. A.1a corresponds to (A.1), Fig. A.1b–e show the program state after iteration 1–4 (at the end of the loop body). The conditional in Line 13 evaluates to `true` in all four iterations. Note that for a different control flow path the pointer $u$ in Fig. A.1 can point to a different sub-tree. The first challenge is to describe Fig. A.1f with recursive predicates (required for convergence of the fix-point calculation). We need an analysis that ensures fix-point convergence and maintains a correct description of the function's the context $tree(root)$, and how the loop statements modify the context. If we were to run our standard loop analysis on the **Reflect tree** example above without the requirement to take the function's context into account, we could ensure convergence of the fix-point iteration with a *garbage* predicate. Since here we want to take the context into account, we could fold the linked list and sub-trees (grey-shaded rectangle and triangles) into a *pls* predicate: $pls(E, F) \iff E \mapsto [\mathtt{u} : u'_1, \mathtt{n} : n'_1] * tree(u'_1) * pls(n'_1, F)$. However, the remaining dark grey tree nodes from *root* to *u* cannot be absorbed in any recursive predicate available in our standard analysis: we cannot use the *tree* predicate because (1) this would mean losing track of the *u* pointer, and (2) the sub-trees would be shared between the *pls* and the *tree* predicate, a fact that is ruled out by the $*$-operator. Note that this problem does not occur in the previous example `filter` because the tree traversal is destructive.

Our first step to solve this problem is to introduce a new predicate to describe the traversed tree segment. We use a *tree segment* predicate to specify structures like the segment from *root* to *u*:

**Definition A.1** (*Tree segment*)

$$
\begin{aligned}
tseg(E, F) \iff & (E \neq F \wedge E \mapsto [\mathtt{l} : t', \mathtt{r} : n'] * tree(t') * tseg(n', F)\ ) \vee \\
& (E \neq F \wedge E \mapsto [\mathtt{l} : n', \mathtt{r} : t'] * tree(t') * tseg(n', F)\ ) \vee \\
& (E = F \wedge emp)
\end{aligned} \tag{A.2}
$$

*i.e. a list segment with an additional pointer to a sub-tree at each node. Each list node connects to its successor either via the left or right pointer field of the original tree node.*

Definition A.1 allows us to describe the segment $tseg(root, u)$. However, we must find a way how *pls* and *tseg* can ‘co-exist’ in a state formula. The formula $tseg(root, u) * pls(s, \mathtt{nil})$ does not express the correlation between *pls* and *tseg*: the nodes of both data structures contain pointers to the sub-trees and these sub-trees are shared by both. The $*$-operator enforces disjointness of these sub-trees, which is wrong in this case. Replacing ‘$*$’ by ‘$\wedge$’ is no viable solution either, because it states potential overlapping of nodes (list nodes and traversed tree nodes) that *are* disjoint in reality. The next section describes a modified analysis that can combine *pls* and *tseg* to correctly describe the program state of our motivating example.

## A.1 Overlaid Sub-Analyses

Our goal is to construct a state formula that allows us to use both the *pls* and *tseg* predicates and correctly express the heaplets shared by these predicates. We borrow and modify a technique developed for overlaid data structures [1] for this purpose. The approach is to split the analysis into two sub-analyses. The separation logic formulae in this analysis are of the form $\psi_C \wedge \phi_L$, where $\psi_C$ generates the partitioning information for the program context and $\phi_L$ describes the state manipulated by the function under test which contains the loop invariant after fix-point convergence. For ease of explanation, we label the context ($C$) analysis with $\psi_C$ and the loop ($L$) analysis with $\phi_L$. The classical conjunction ('$\wedge$') between these sub-analyses allows sharing, i.e. both analyses *can* describe the same heap-allocated objects (in contrast to predicates connected by '$*$' which strictly requires that the described objects be disjoint). The formulae in $\psi_C$ and $\phi_L$ focus on different objectives and parts of the program, but the combined formula $\psi_C \wedge \phi_L$ is always a valid and accurate assertion for the program state. By dividing the problem into $\psi_C$ and $\phi_L$ we are able to simultaneously use the *tseg* and *pls* predicates. As in [1], we transfer information from the loop analysis to the context analysis about the effect of the loop body on the context. A difference is that we need to transfer information only in one direction: from the loop analysis $\phi_L$ to the context analysis $\psi_C$. As described above, the call-site predicate of the function `reflectTree` is *tree*(*root*). At the start, we include this predicate in both sub-analyses and enter the symbolic execution of the function with the state formula:

$$\big(tree(root)_\alpha * \texttt{true}_\beta\big) \wedge \big(tree(root)_\alpha * \texttt{true}_\alpha\big) \tag{A.3}$$

In the following discussion, we call the left (right) side of the conjunction the left (right) hand side. The state formula on the right hand side evolve in the same way as the loop analysis in Sect. 4.2 (with a slight modification discussed below). The left hand side receives state updates that are transferred from the loop analysis to the context analysis; in the end it contains the partition information provided to the program context. We introduce the `true` predicate on both sides. This assertion always holds and allows us to absorb ('remove') predicates that are not relevant for the respective sub-analysis without falsifying the overall state assertion. Absorbing predicates in `true` can be viewed as 'weakening' the formula, i.e. making it less precise. The idea, however, is that the 'missing' information in one sub-analysis is preserved on the other side and vice versa. As in [1], we assign region variables $\alpha$ and $\beta$ to the predicates. We use them to express what predicates belong to the context data and what predicates belong to heaplets that are only used during function execution. The predicate $tree(root)_\alpha$ states that the addresses of the tree nodes form the set $\alpha$. The region variables can be straightforwardly integrated in our heaplet label sets in Sect. 4.2. However, $\alpha$ and $\beta$ should not be confused with heap footprint labels (we omit footprint labels in the following equations for ease of readability). In the following discussion, we repeat parts of the description of the cut-point insertion and fix-point calculation for the extended analysis. In the case of the `reflectTree` example, we require one loop iteration to be peeled off in order to prove a valid partitioning. We execute the loop preamble and obtain:

$$\begin{aligned}&\big(tree(root)_\alpha * \texttt{true}_\beta\big) \wedge \\ &\big(tree(root)_\alpha * s \mapsto [\texttt{u}: root, \texttt{n}: \texttt{nil}]_\beta * \texttt{true}_\alpha\big)\end{aligned} \tag{A.4}$$

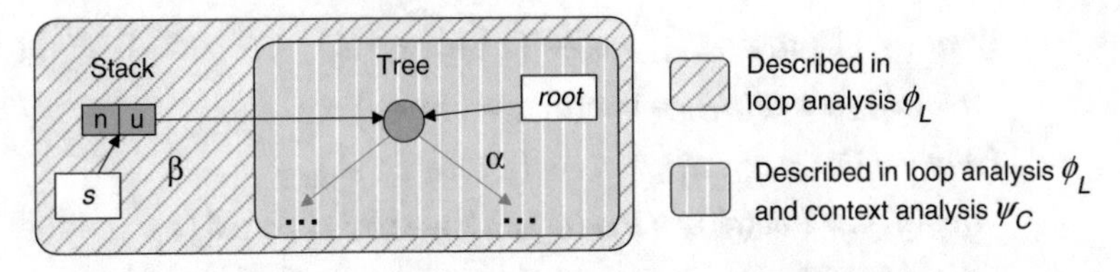


**Fig. A.2** Program state described by the two sub-analyses $\phi_L$ and $\psi_C$ in (A.4)

We have not accessed the tree, i.e. the context, yet. The linked list node has been created by the preamble, i.e. by the function under test itself and is therefore automatically assigned to region $\beta$. Figure A.2 shows the program state described by the two sub-analyses in (A.4). For ease of explanation, we assume that the sequence of evaluations of the conditional in Line 13 is the same as in Fig. A.1. We unroll the first iteration and, at the end of the loop body, we get:

$$\begin{aligned}&\big(tree(root)_\alpha * \mathtt{true}_\beta\big) \wedge \\ &\big(root = u \wedge u \mapsto [\mathtt{l} : u_0', \mathtt{r} : u_1']_\alpha * tree(u_0')_\alpha * tree(u_1')_\alpha * \\ &\quad s \mapsto [\mathtt{u} : u_0', \mathtt{n} : s_0']_\beta * s_0' \mapsto [\mathtt{u} : u_1', \mathtt{n} : \mathtt{nil}]_\beta * \mathtt{true}_\alpha\big)\end{aligned} \tag{A.5}$$

The right hand side of (A.5) shows that we have accessed the context because predicates of region $\alpha$ have been modified (the root record was accessed by a program command): $tree(root)_\alpha$ was transformed into $root = u \wedge u \mapsto [\mathtt{l} : u_0', \mathtt{r} : u_1']_\alpha * tree(u_0')_\alpha * tree(u_1')_\alpha$. However, the left hand side still has no knowledge of this change. We transfer this information to the context sub-analysis by applying the same state transition in $\psi_C$:

$$\begin{aligned}&\big(root = u \wedge u \mapsto [\mathtt{l} : u_0', \mathtt{r} : u_1']_\alpha * \\ &\quad tree(u_0')_\alpha * tree(u_1')_\alpha * \mathtt{true}_\beta\big) \wedge \\ &\big(root = u \wedge u \mapsto [\mathtt{l} : u_0', \mathtt{r} : u_1']_\alpha * tree(u_0')_\alpha * tree(u_1')_\alpha * \\ &\quad s \mapsto [\mathtt{u} : u_0', \mathtt{n} : s_0']_\beta * s_0' \mapsto [\mathtt{u} : u_1', \mathtt{n} : \mathtt{nil}]_\beta * \mathtt{true}_\alpha\big)\end{aligned} \tag{A.6}$$

The cut-point insertion now finds a valid cut-point pair $s$ and $s_b = s_0'$ (it inserts $s_b$). At this point, our tool submits the instrumented program state to the fix-point calculation. After the fix-point calculation executed the second iteration, at the end of the loop body and after all sub-analysis synchronisation, we obtain ($s_b$ inserted for $s_0'$):

$$\begin{aligned}&(root \mapsto [\mathtt{l}: u, \mathtt{r}: u_1']_\alpha * u \mapsto [\mathtt{l}: u_2', \mathtt{r}: u_3']_\alpha *\\&tree(u_2')_\alpha * tree(u_3')_\alpha * tree(u_1')_\alpha * \mathtt{true}_\beta) \wedge\\&(root \mapsto [\mathtt{l}: u, \mathtt{r}: u_1']_\alpha * u \mapsto [\mathtt{l}: u_2', \mathtt{r}: u_3']_\alpha *\\&tree(u_2')_\alpha * tree(u_3')_\alpha * tree(u_1')_\alpha * s \mapsto [\mathtt{u}: u_2', \mathtt{n}: s_1']_\beta *\\&s_1' \mapsto [\mathtt{u}: u_3', \mathtt{n}: s_b]_\beta * s_b \mapsto [\mathtt{u}: u_1', \mathtt{n}: \mathtt{nil}]_\beta * \mathtt{true}_\alpha)\end{aligned} \tag{A.7}$$

In contrast to the 'standard' symbolic execution of loop iterations, we can now steer the two sub-analyses towards a different objectives. The sub-analysis $\phi_L$ focuses on the fix-point calculation for the loop only and hence needs to consider only those program variables that are 'used' by the loop. On the other hand, the sub-analysis $\psi_C$ focuses on the context predicates only and must consider program variables whose scope extends to the program context. Before launching the analysis, partition the set of program variables ($PV$) in the two groups: loop variables ($LV$) and context variables ($CV$). We obtain the former group by running a *definition-usage* (DEF-USE) analysis, a standard LLVM analysis, that lists all pointer variables read or written to within the scope of the loop. In this case, $LV = \{s, u, l, r\}$. The latter group consists of pointer variables that are not declared within the function body. In this case, $CV = \{root\}$. We use this information about program variables to absorb predicates that arise in the loop analysis in the `true` assertion. To this end, the analysis applies 'absorption rules' that are specifically designed for the sub-analysis $\phi_L$. The following example shows a rule for a tree node predicate. The rule is analogously defined for nodes in *ls* and *pls* predicates.

**Definition A.2** (*Absorption rule for a tree node predicate in* $\phi_L$)

$$\frac{E \in PV \wedge E \notin LV \wedge E \in CV}{E \mapsto [\mathtt{l}: F, \mathtt{r}: G]_\alpha \rightsquigarrow \mathtt{true}_\alpha} \tag{A.8}$$

*i.e. the predicate is absorbed if the predicate describes context data (region* $\alpha$*), and the pointer expression E referencing it is a program variable, and is not read or written within the loop under test, and is declared in the program context. Note that a predicate cannot be recovered after it has been absorbed by the* `true` *assertion.*

The predicate $root \mapsto [\mathtt{l}: u, \mathtt{r}: u_1']_\alpha$ on the right hand side satisfies the condition of Definition A.2 and is absorbed by $\mathtt{true}_\alpha$:

$$\begin{aligned}&(root \mapsto [\mathtt{l}: u, \mathtt{r}: u_1']_\alpha * u \mapsto [\mathtt{l}: u_2', \mathtt{r}: u_3']_\alpha *\\&tree(u_2')_\alpha * tree(u_3')_\alpha * tree(u_1')_\alpha * \mathtt{true}_\beta) \wedge\\&(u \mapsto [\mathtt{l}: u_2', \mathtt{r}: u_3']_\alpha *\\&tree(u_2')_\alpha * tree(u_3')_\alpha * tree(u_1')_\alpha * s \mapsto [\mathtt{u}: u_2', \mathtt{n}: s_1']_\beta *\\&s_1' \mapsto [\mathtt{u}: u_3', \mathtt{n}: s_b]_\beta * s_b \mapsto [\mathtt{u}: u_1', \mathtt{n}: \mathtt{nil}]_\beta * \mathtt{true}_\alpha)\end{aligned} \tag{A.9}$$

The fix-point calculation proceeds in the same way as described in the previous section: symbolic execution of loop iterations and abstraction. The latter requires a small modification which is explained in the next section.

## A.2 Abstraction and Fix-point Convergence

The fix-point calculation for the extended analysis uses a slightly modified set of abstraction rules. When folding singleton heaplets into recursive predicates, the new abstraction rules ensure that the information about region variables $\alpha$ and $\beta$ is not lost. To this end, the region variables attached to predicates are assigned in the same order to the recursive predicate. We implement this behaviour in a modification to our standard set of abstraction rules. The following example shows a folding operation using the new abstraction rule for the *pls* predicate:

$$E \mapsto [\mathtt{u} : u_1', \mathtt{n} : n_1']_{r1} * tree(u_1')_{r2} * n_1' \mapsto [\mathtt{u} : u_2', \mathtt{n} : F]_{r1} * tree(u_2')_{r2} * \rightsquigarrow pls(E, F)_{r1,r2}$$

The region variables $r1$ and $r2$ are placeholders and can each take the value $\alpha$ or $\beta$. Note that we omit the additional footprint labels here for clarity. When a node in the list is accessed by the symbolic execution, $pls(E, F)_{r1,r2}$ unfolds to

$$E \mapsto [\mathtt{u} : u_1', \mathtt{n} : n_1']_{r1} * tree(u_1')_{r2} * pls(n_1', F)_{r1,r2}$$

Abstraction is performed in both sub-analyses $\psi_C$ and $\phi_L$ which are both required to converge in order to generate a valid proof. The abstraction rules for both analyses only differ in the way heap footprint labels are treated: In $\psi_C$, we do not merge predicates with different footprint labels, i.e. $\langle u \mapsto [l : u_1', r : u_2']_\alpha \rangle_{\{\}} * \langle tree(u_1')_\alpha \rangle_{\{a\}} * \langle tree(u_2')_\alpha \rangle_{\{b\}}$ gets folded into $\langle tree(u)_\alpha \rangle_{\{a,b\}}$ in the loop analysis $\phi_L$, but not in the context analysis $\psi_C$. This is necessary because we may lose the partitioning information for the context otherwise. In our `reflectTree` example, we reach the following loop-invariant state:

$$\begin{aligned}
&\big(\langle root \mapsto [\mathtt{l} : u_6', \mathtt{r} : u_7']_\alpha \rangle_{\{\}} * \langle tseg(u_6', u)_\alpha \rangle_{\{a\}} * \\
&\langle tree(u)_\alpha \rangle_{\{a\}} * \langle tree(u_7')_\alpha \rangle_{\{a\}} * \mathtt{true}_\beta \vee \\
&\langle root \mapsto [\mathtt{l} : u_6', \mathtt{r} : u_7']_\alpha \rangle_{\{\}} * \langle tseg(u_7', u)_\alpha \rangle_{\{b\}} * \\
&\langle tree(u)_\alpha \rangle_{\{b\}} * \langle tree(u_6')_\alpha \rangle_{\{b\}} * \mathtt{true}_\beta\big) \wedge (\phi)
\end{aligned} \tag{A.10}$$

Due to space limitations, we write out the context assertion $\psi_C$ only. At this point, the fix-point calculation terminates with the usual partition label assignment and our analysis 'leaves' the loop. The left hand side of (A.10) contains the information we provide to the program context. Because the program variable $u$ is local to the function `reflectTree` and is not defined outside its scope ($u \notin CV$), the analysis replaces it with a fresh primed variable $v_1'$. This automatically triggers a last abstraction step in the context assertion after the fix-point calculation terminated and results in:

$$\begin{aligned}&\big(\langle root \mapsto [\mathtt{l} : u_6', \mathtt{r} : u_7']_\alpha\rangle_{\{\}} * \\ &\langle tree(u_6')_\alpha\rangle_{\{a\}} * \langle tree(u_7')_\alpha\rangle_{\{b\}} * \mathtt{true}_\beta\big) \wedge (\phi)\end{aligned} \tag{A.11}$$

The disjunctive clauses in (A.10) became equivalent after the abstraction and were automatically conjoined by the tool. As in Sect. 4.2, (A.11) tells us that the heap accessed by the loop can be partitioned into two disjoint regions labelled $a$ and $b$. Furthermore, it tells us how the algorithm partitioned the heap that is accessed not only by the function itself but also by other parts of the program (the context). For example, the source code transformation in a function that builds the tree data structure can use this context information to assign the correct memory bank to each partition.

## Reference

1. O. Lee, H. Yang, R. Petersen, Program analysis for overlaid data structures, in *Proceedings of the International Conference on Computer Aided Verification* (2011), pp. 592–608

Printed in the United States
By Bookmasters